Statistical Methods for
Speech Recognition

Language, Speech, and Communication

Statistical Methods for Speech Recognition

Frederick Jelinek

The MIT Press
Cambridge, Massachusetts
London, England

This book was set in Times New Roman by Asco Typesetters, Hong Kong.

Printed and bound in the United States of America.

Library of Congress Cataloging-in-Publication Data

Jelinek, Frederick, 1932–
  Statistical methods for speech recognition / Frederick Jelinek.
    p.   cm. — (Language, speech, and communication)
  Includes bibliographical references and index.
  ISBN 0-262-10066-5 (alk. paper)
  1. Automatic speech recognition—Statistical methods. I. Title.
II. Series.
TK7895.S65J45   1998   97-34860
006.4′54—dc21   CIP

10 9 8 7 6 5

To Milena

In memory of my parents

MUDr. Vilém Jelínek
★ 29 April 1898, Ivančice
† 11 May 1945, Terezín (Theresienstadt)

Trude Jelinek
★ 21 May 1909, Zürich
† 16 February 1996, Zürich

Contents

Chapter 7

Contents

Chapter 12

Chapter 13

Preface

The origin of this book is a set of notes for a course I taught in 1991–92 at the Czech Technical University. I revised them extensively for students at the California Institute of Technology (spring 1993) and Johns Hopkins University. By 1994, the handouts had started to be called chapters.

This is not a textbook of speech recognition, but only of its statistical aspects. The material reflects my own research interests: There is no discussion of signal processing, which is so crucial to good performance. I am fascinated by the idea that while system structure and parametrization should come from intuitive understanding of the process, the parameter values are best extracted from actual data. This book tells how to extract such information when data is sparse—the pervasive reality with which any researcher must come to terms.

The text concentrates on those basic statistical ideas that have proven so fruitful in speech recognition: hidden Markov models, data clustering, smoothing of probability distributions, the decision tree method of equivalence classification, the use of information measures as goodness criteria, and maximum entropy probability estimation. The aim is clarity, conciseness, and a unified point of view.

The book is intended to teach principles that can be applied in a variety of circumstances, not to give detailed recipes for system development. In many instances the approach presented is a simplification of actual practice. In this text for instance, except for chapter 9 where generality is assumed, the input of acoustic data to the recognizer's linguistic decoder is discrete. In any case, experience shows that every application requires a painstaking working out, that basically correct ideas must always be modified to fit the particular circumstances.

Our treatment is not exhaustive, but each chapter contains a section called "Additional Reading" that points to articles revealing more details, refinements, and experimental results. Here an apology is surely in order: The selection of references in these sections was not made with any pretense of fairness to the many important contributors to the field. The author has simply called the reader's attention to material with which he is familiar that seemed to complement the chapter's contents. The many regrettable omissions are due to the author's vast ignorance of the literature. No slights were intended.

There are few explicit prerequisites to understanding the book's contents. Some familiarity with probability, college level mathematics, and (mainly) maturity are required. We progress step by step, trying to be self-contained and appealing to common sense. No advanced theorems are invoked, not even the central limit theorem or the theory of matrices with nonnegative elements.

The book tries to introduce methods heuristically, sometimes delaying mathematical proofs, often eschewing them altogether. The reader will notice that few claims for optimality are made. The practical experience of speech recognition shows that adequacy should be sought, and that "greedy" approaches pay off. Except for the Additional Reading sections, the references are mainly confined to essential explanations from other fields and to the pioneering articles from our own. Unified notation is strived for: **boldface** represents vectors, strings, and sequences; $\mathscr{CURSIVE}$ represents sets.

The plan of this text is to present first (chapters 1 through 5) each basic component of a large vocabulary speech recognizer (acoustic model, language model, hypothesis search) and devote the rest of the book (chapters 6 through 15) to refinements.

In particular, chapter 1 introduces the fundamental statistical formulation of the speech recognition problem. Chapter 2 discusses hidden Markov models in some detail, including the Baum algorithm as derived from a heuristic argument. Then we are ready to introduce acoustic modeling in chapter 3. Chapter 4 treats basic language modeling, and chapter 5 completes the design of a rudimentary large vocabulary speech recognizer by describing a Viterbi-based hypothesis search.

We are then in a position to become more sophisticated. In chapter 6 we conclude the discussion of the hypothesis search by describing the multistack algorithm and a version of the fast match. Since further exposition will often require information measures of goodness, we dedicate chapter 7 to a self-contained survey of basic information theory. It is then natural to address the problems of comparative complexity of recognition tasks and of the quality-of-language models, introducing the concept of perplexity that is central to chapter 8. We can then clean up and prove the Baum algorithm in considerable generality, including continuous feature vectors, tied parameters, etc. Chapter 9 therefore presents the expectation-maximization algorithm.

Chapter 10 describes the decision tree method of equivalence classification through the vehicle of language modeling. The next two chapters apply decision trees to phonetics. Chapter 11 treats the problem of phonetic base form creation from spelling and speech data, and chapter 12 is concerned with allophonic variation, handling the triphone concept, among others. Chapter 13 focuses on estimating probability distributions satisfying imposed linear constraints. This problem is solved by the maximum entropy method, which we again explain in a language model setting. Chapter 14 applies the approach to three different aspects of language modeling. Up to this point, all probability estimates have been closely tied to relative frequencies of observed data. Chapter 15 completes the book by discussing other alternatives, including the well-known Good-Turing estimation.

Concluding this preface, I would like to acknowledge what will be obvious to those familiar with the field: Members of the IBM Continuous Speech Recognition Group, of which I am proud to have been part, have pioneered most of the methods presented here. The foundation was laid in those heady, long-gone years when wise management supported us with generous computing resources, left us free to work as we thought fit, and did not yet push slogans like "customer oriented" or "market driven." John Cocke was our patron, and noontime walks with Lalit Bahl, Bob Mercer, and (somewhat later and for too short a time) Jim Baker provided continual stimulation.

Finally, I would like to give special thanks to Kimberly Shiring, Anne Millman Storck, Timothy Edwards, and Asela Gunawardana for the crucial help they have given me in the preparation of the manuscript of this book.

Chapter 1

The Speech Recognition Problem

1.1 Introduction

A speech recognizer is a device that automatically transcribes speech into text. It can be thought of as a voice-actuated "typewriter" in which a computer program carries out the transcription and the transcribed text appears on a workstation display. The recognizer is usually based on some finite vocabulary that restricts the words that can be "printed" out. Until we state otherwise, the designation *word* denotes a *word form* defined by its spelling. Two differently spelled inflections or derivations of the same stem are considered different words (e.g., *table* and *tables*). Homographs having different parts of speech (e.g., *absent* [VERB] and *absent* [ADJECTIVE]) or meanings (e.g., *bank* [FINANCIAL INSTITUTION] and *bank* [OF A RIVER]) constitute the same word.

Figure 1.1, the speech waveform corresponding to an utterance of the chess move "*Bishop moves to king knight five*," illustrates our problem. The waveform was produced at Stanford University in the late 1960s in a speech recognition project undertaken by Raj Reddy [1].

Because the field was then in its infancy, Reddy tried to give himself all the conceivable but fair advantages. Recognition of spoken chess moves could be based on a small vocabulary, a restricted syntax, and a relatively complete world knowledge: The system knew which chess moves were legal, so that, for instance, a recognition hypothesis corresponding to a move of a piece to a square occupied by another piece of the same color could immediately be rejected. The speech was recorded by a very good microphone in a quiet environment, and the system was adjusted (as well as the state of the art allowed) to the speaker.

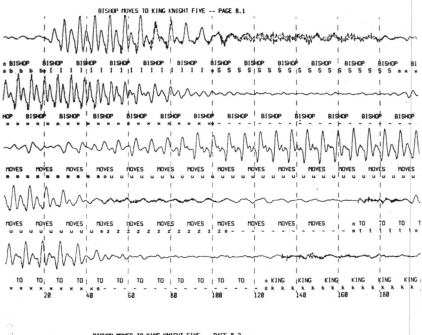

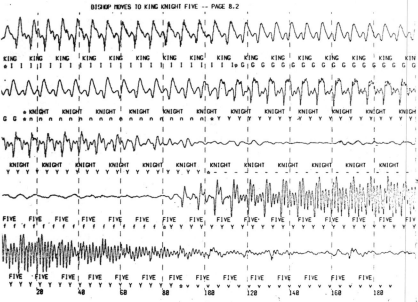

Figure 1.1
The sentence "Bishop moves to king knight five" aligned with its speech waveform

In early design strategy of the field, a recognizer would segment the speech into successive phones (basic pronunciation units [2]), then identify the particular phones corresponding to the segments, and finally transcribe the recognized phone strings into an English text.

Figure 1.1 aligns the speech waveform with the spoken words. Inspecting it, we can see that although the boundary between different speech events cannot be accurately placed, distinct events certainly appear in succession. So even though the boundaries are fuzzy, most of the segments seem to contain repeating nuclei that are candidates for recognition. Although this does not hold for stops[1] like b (see the beginning of BISHOP) or k (beginning of KING), it does seem to hold for vowels such as i (BISHOP and KING) or u (the first vowel of MOVES). But inspecting figure 1.1 more closely, we encounter a very big problem: The i of KING looks much more like the u of MOVES than it does like the i of BISHOP! So visually similar waveforms do not necessarily indicate perceptually similar sounds.

Actually, context is involved here, and the apparent waveform similarities are due to the influence of the nasality of the sound of ng that follows the i and of the sound of m that precedes the u sound.

Having now indicated an aspect of why speech recognition is not an easy task, we are ready to begin its serious consideration. In this chapter we will formulate the large vocabulary[2] speech recognition problem mathematically, which will result in a recognizer's natural breakup into its components. The chapter will conclude with the first example of self-organization from data: the basic vector quantization algorithm [11] that can be used to transform the speech signal into a sequence of symbols from a relatively limited (and automatically selected!) alphabet.

---

1. Stops, also called plosives, are sounds that consist of two parts, a stop portion and a release portion. English stops are b, d, g, k, p, t. [2]

2. Although no hard and fast distinction between small and large vocabulary tasks exists, here are some examples of each:

• Small vocabulary: recognition of digits, yes-no answers to questions, inventory control, etc.

• Large vocabulary: text creation by dictation, transcription of e-mail, transcription of telephone conversations, text creation for the handicapped.

Small vocabulary tasks are of great economic importance. They are just not the subject of this book.

1.2 A Mathematical Formulation

To discuss the problem of speech recognizer design, we need its mathematical formulation. A precise statement of the problem leads directly to
a fruitful decomposition into easier to treat subproblems. Our approach is
statistical,[3] so the formulation will involve probabilities. Here it is [3] [4]:

Let $\mathbf{A}$ denote the acoustic evidence (data) on the basis of which the recognizer will make its decision about which words were spoken. Because
we are dealing with digital computers, then without loss of generality we
may assume that $\mathbf{A}$ is a sequence of symbols taken from some (possible
very large) alphabet $\mathscr{A}$:

$$\mathbf{A} = a_1, a_2, \ldots, a_m \qquad a_i \in \mathscr{A} \tag{1}$$

The symbols a_i can be thought of as having been generated in time, as
indicated by the index i.

Let

$$\mathbf{W} = w_1, w_2, \ldots, w_n \qquad w_i \in \mathscr{V} \tag{2}$$

denote a string of n words, each belonging to a fixed and known vocabulary $\mathscr{V}$.

If $P(\mathbf{W}|\mathbf{A})$ denotes the probability that the words $\mathbf{W}$ were spoken, given
that the evidence $\mathbf{A}$ was observed, then the recognizer should decide in
favor of a word string $\hat{\mathbf{W}}$ satisfying

$$\hat{\mathbf{W}} = \arg \max_{\mathbf{W}} P(\mathbf{W}|\mathbf{A}) \tag{3}$$

That is, the recognizer will pick the most likely word string given the
observed acoustic evidence.

Of course, underlying the target formula (3) is the tacit assumption that
all words of a message are equally important to the user, that is, that
misrecognition does not carry a different penalty depending on which
word was misrecognized. Under this philosophy the warning *"Fire!"*

---

3. No advanced results of probability or statistical theory will be used in this self-
contained text. The student is required simply to be comfortable with statistical
concepts and be able to manipulate them intuitively. So although nothing in this
text presumes more than the knowledge of the first four chapters of a book like [5],
the required sophistication can probably be gained only by completing an entire
course.

carries no more importance in a crowded theater than an innocuous commercial announcement. But let that possible criticism pass.[4]

The well known Bayes' formula of probability theory allows us to rewrite the right-hand side probability of (3) as

$$P(\mathbf{W}|\mathbf{A}) = \frac{P(\mathbf{W})P(\mathbf{A}|\mathbf{W})}{P(\mathbf{A})} \tag{4}$$

where $P(\mathbf{W})$ is the probability that the word string $\mathbf{W}$ will be uttered, $P(\mathbf{A}|\mathbf{W})$ is the probability that when the speaker says $\mathbf{W}$ the acoustic evidence $\mathbf{A}$ will be observed, and $P(\mathbf{A})$ is the average probability that $\mathbf{A}$ will be observed. That is,

$$P(\mathbf{A}) = \sum_{\mathbf{W}'} P(\mathbf{W}')P(\mathbf{A}|\mathbf{W}') \tag{5}$$

Since the maximization in (3) is carried out with the variable $\mathbf{A}$ fixed (there is no other acoustic data save the one we are given), it follows from (3) and (4) that the recognizer's aim is to find the word string $\hat{\mathbf{W}}$ that maximizes the product $P(\mathbf{W})P(\mathbf{A}|\mathbf{W})$, that is, it satisfies

$$\hat{\mathbf{W}} = \arg \max_{\mathbf{W}} P(\mathbf{W})P(\mathbf{A}|\mathbf{W}) \tag{6}$$

1.3   Components of a Speech Recognizer

Formula (6) determines what processes and components are of concern in the design of a speech recognizer.

1.3.1   Acoustic Processing

First, it is necessary to decide what acoustic data $\mathbf{A}$ will be observed. That is, one needs to decide on a *front end* that will transform the pressure waveform (which is what sound is) into the symbols a_i with which the recognizer will deal. So in principle, this front end includes a microphone whose output is an electric signal, a means of sampling that signal, and a manner of processing the resulting sequence of samples.

---

4. Strictly speaking, formula (3) is appropriate only if we are after a perfect transcription of the utterance, that is, if one error is as bad as many. Were we to accept that errors are inevitable (which they certainly are) and aim explicitly at minimizing their number, a much more complex formula would be required. So our formula only approximates (it turns out very fruitfully) what we are intuitively after.

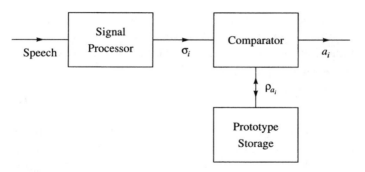

Figure 1.2
A schematic diagram of a recognizer front end (acoustic processor)

In this book, dedicated to statistical models of speech recognition, we will not address the front-end problem, except in the penultimate section of this chapter.[5] For the present, we will assume that the alphabet $\mathscr{A}$ is simply given. Those interested in front-end design should consult the many books and articles that thoroughly discuss *signal processing* [6] [7]. For the reader to gain some idea of what might be involved, however, we present in figure 1.2 a schematic diagram of a rudimentary front end (*acoustic processor*).

We can think of the *signal processor* as a device that at regular intervals of time (e.g., a hundred times per second[6]) generates real-valued vectors σ_i. The components of σ_i could be sample values of outputs of band-pass filters applied to the signal coming out of the microphone.

The *prototype storage* contains a set of vector prototypes $\mathscr{R} = \{\rho_1, \rho_2, \ldots, \rho_K\}$ of the same kind as σ_i. The *comparator* finds the closest element of $\mathscr{R}$ to σ_i, and the index of that element *is* the acoustic symbol a_i. To be precise,

$$\hat{J} = \arg \min_{j=1}^{K} d(\sigma_i, \rho_j) \tag{7}$$

and

5. Obviously, good signal processing is crucial and is the subject of intensive research. This book is about extracting information from the processed signal. Bad processing means loss of information: There is less of it to extract.

6. This happens to be the prevalent standard in our field.

$$a_i = \hat{J} \tag{8}$$

In (7), $d(\ ,\ )$ denotes a suitable distance function.[7]

In the penultimate section of this chapter we introduce a simple method, called *vector quantization*, that can derive the prototype set $\mathscr{R} = \{\rho_1, \rho_2, \ldots, \rho_K\}$ from speech data. There, the intuition behind the output symbol selection rule (7) and (8) will become apparent.

In most state-of-the-art large vocabulary recognition systems the comparator of figure 1.2 is omitted and the rest of the recognizer handles directly the signal processor outputs σ_i. They then constitute the observable symbols a_i. To introduce the relevant methods in the simplest setting, however, we will assume an acoustic symbol alphabet size of the order of hundreds (the size 200 is very common), which will still allow us to deal with the essence of the problem. We will generalize certain important results to "continuous" vector spaces (for signal processor outputs σ) in chapter 9. The added complication is that statistical estimation for vector spaces requires parametric methods. Section 2.9.1 of the next chapter will briefly introduce an appropriate model.

1.3.2 Acoustic Modeling

Returning now to formula (6), the recognizer needs to be able to determine the value $P(\mathbf{A}|\mathbf{W})$ of the probability that when the speaker uttered the word sequence $\mathbf{W}$ the acoustic processor produced the data $\mathbf{A}$. Since this number must be made available for all possible pairings of $\mathbf{W}$ with $\mathbf{A}$, it follows that it must be *computable "on the fly."* The number of different possible values of $\mathbf{A}$ and $\mathbf{W}$ is just too large to permit a lookup.

Thus to compute $P(\mathbf{A}|\mathbf{W})$ we need a statistical *acoustic model* of the speaker's interaction with the acoustic processor. The total process we are modeling involves the way the speaker *pronounces* the words of $\mathbf{W}$, the ambience (room noise, reverberation, etc.), the microphone placement and characteristics, and the acoustic processing performed by the front end.

The usual acoustic model employed in speech recognizers, the *hidden Markov model*, will be discussed in the next chapters. Other models are possible, for instance those based on *artificial neural networks* [8] or on *dynamic time warping* [9]. These methods are not treated in this book.

---

7. A very adequate distance is Euclidean:

$$d(\mathbf{x}, \mathbf{y}) \doteq \sqrt{\sum_i (x_i - y_i)^2}$$

1.3.3 Language Modeling

Formula (6) further requires that we be able to compute for every word string $\mathbf{W}$ the a priori probability $P(\mathbf{W})$ that the speaker wishes to utter $\mathbf{W}$. Bayes' formula allows many decompositions of $P(\mathbf{W})$, but because the recognizer "naturally" wishes to convey the text in the sequence in which it was spoken, we will use the decomposition

$$P(\mathbf{W}) = \prod_{i=1}^{n} P(w_i|w_1, \ldots, w_{i-1}) \tag{9}$$

The recognizer must thus be able to determine estimates of the probabilities $P(w_i|w_1, \ldots, w_{i-1})$. We use the term *estimate* on purpose, because even for moderate values of i and vocabularies of reasonable size, the probability $P(w_i|w_1, \ldots, w_{i-1})$ has just too many arguments. In fact, if $|\mathcal{V}|$ denotes the size of the vocabulary, then for $|\mathcal{V}| = 20,000$ and $i = 3$, the number of arguments is 8×10^{12}.

It is, of course, absurd to think that the speaker's choice of his i^{th} word depends on the entire *history* $w_1, \ldots, w_{i-1}$ of all of his previous speech. It is therefore natural that for purposes of the choice of w_i, the history be put into *equivalence classes* $\Phi(w_1, \ldots, w_{i-1})$. Thus in reality formula (9) becomes

$$P(\mathbf{W}) = \prod_{i=1}^{n} P(w_i|\Phi(w_1, \ldots, w_{i-1})) \tag{10}$$

and the art of *language modeling* consists of determining the appropriate equivalence classification Φ and a method of estimating the probabilities $P(w_i|\Phi(w_1, \ldots, w_{i-1}))$.

It is worth stressing that the language model used should depend on the use to which the recognizer will be put. The transcription of dictated radiological reports requires different language models than the writing of movie reviews. If text is to be produced, then the language model may reasonably be constructed by processing examples of corresponding written materials. It will then depend on text only and not in any way on speech.

1.3.4 Hypothesis Search

Finally, to find the desired transcription $\hat{\mathbf{W}}$ of the acoustic data $\mathbf{A}$ by formula (6), we must search over all possible word strings $\mathbf{W}$ to find the maximizing one. This search cannot be conducted by brute force: The space of $\mathbf{W}$s is astronomically large.

A parsimonious hypothesis search is needed that will not even consider the overwhelming number of possible candidates **W** and will examine only those word strings in some way suggested by the acoustics **A**. We will devote chapters 5 and 6 to showing two ways to proceed.

1.3.5 The Source-Channel Model of Speech Recognition

Our formulation leads to the schematic diagram of figure 1.3. The human speaker is shown as consisting of two parts: The source of the communication is his mind, which specifies the words **W** that will be pronounced by his vocal apparatus, the speech producer. The recognizer also consists of two parts, the acoustic processor and the linguistic decoder, the latter containing the acoustic and language models and the hypothesis search algorithm. Figure 1.3 embeds the process into the communication theory [10] framework: the source (modeled by the language model), the "noisy" channel (consisting of the tandem of the speech producer and the acoustic processor and modeled by the acoustic model), and the *linguistic decoder*.

The only difference between our communication situation and the standard one is that in the standard, the system designer can introduce an encoder and a modulator (signal generator) between the source and the channel. We must make do with the coder-modulator that evolution has bequeathed to us: human language and speech.

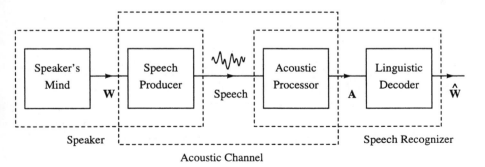

Figure 1.3
Source-channel model of speech recognition

1.4 About This Book

The previous section revealed the topics with which speech recognition research must be concerned. This book investigates methods that would accomplish the tasks outlined in sections 1.3.2 through 1.3.4. The headings of the above sections use the expression *modeling* because the required values of $P(\mathbf{A}|\mathbf{W})$ and $P(\mathbf{W})$ can only be produced by evaluating probabilities of output strings generated by abstract models of corresponding processes. These models will have no more than a mathematical reality. No claims whatever can conceivably be made about their relation to humans' actual speech production or recognition.

For all practical purposes the rest of the book is devoted to the structure of models of progressively higher sophistication, to their parameters, and to the problem of estimating the parameter values from actual speech data.

The plan of this text is to present first (chapters 1 through 5) each basic component of a large vocabulary speech recognizer (acoustic model, language model, hypothesis search) and devote the rest of the book (chapters 6 through 15) to refinements.

1.5 Vector Quantization

We conclude this chapter by introducing a simple method that can be used to find appropriate vector prototypes $\mathcal{R} = \{\rho_1, \rho_2, \ldots, \rho_K\}$ for the prototype storage of the acoustic processor of figure 1.2 (see the discussion of section 1.3.1) [11].

Consider the real vectors σ_i put out periodically by the signal processor. If their dimension is L, then they can be regarded as points in the real L-dimensional space. As speech is input to the signal processor, the space is being occupied by the points σ_i. If speech can be represented as a succession of *phones* [2] produced by the speaker (see the pronunciation specification of words in any dictionary), then the points σ_i will *cluster* in regions of the L-space characteristic of the particular phones the speaker produced. Therefore, if the prototypes $\rho_j \in \mathcal{R}$ are selected to be the centers of the σ-clusters, then the nearest prototype to an observed σ_i will be an appropriate *representative* of σ_i. This in fact is the idea behind the output symbol selection rule (7) and (8).

Vector quantization is a simple and effective method for finding cluster centers. Along with the rule (7) it presupposes a distance measure $d(\ ,\ )$ between points of the L-space. It turns out that a very adequate measure is the Euclidean distance

$$d(\sigma,\rho) = \sqrt{\sum_{l=1}^{L}(\sigma(l) - \rho(l))^2} \tag{11}$$

where $\sigma(l)$ and $\rho(l)$ denote the lth components of the vectors σ and ρ.

The Vector Quantization Algorithm[8]

1. Select K, the number of clusters.
2. Send speech through the signal processor (see figure 1.2) obtaining vectors σ_i for $i = 1, 2, \ldots, N$ (the total amount of speech, proportional to N, must be judiciously chosen).[9]
3. Select uniformly at random K initial candidate cluster centers ρ_j^0 from among the speech vectors $\{\sigma_1, \sigma_2, \ldots, \sigma_N\}$, that is,

$$P\{\rho_j^0 = \sigma_i\} = \frac{1}{N}$$

4. Partition $\{\sigma_1, \sigma_2, \ldots, \sigma_N\}$ into K regions $\mathscr{S}_j^0$ comprising all σ_i nearer to ρ_j^0 than to any other ρ_h^0, $h \neq j$. That is,

$$\sigma_i \in \mathscr{S}_j^0 \quad \text{if } d(\sigma_i, \rho_j^0) \leq d(\sigma_i, \rho_h^0) \text{ for all } h$$

5. Find the *center of gravity* ρ_j^1 of each collection $\mathscr{S}_j^0$. That is,

$$\rho_j^1 = \arg\min_{\rho} \sum_{\sigma_i \in \mathscr{S}_j^0} d(\sigma_i, \rho)$$

6. Repartition $\{\sigma_1, \sigma_2, \ldots, \sigma_N\}$ into K regions $\mathscr{S}_j^1$ comprising all σ_i nearer to ρ_j^1 than to any other ρ_h^1, $h \neq j$.
7. Find the center of gravity ρ_j^2 of each collection $\mathscr{S}_j^1$.
8. And so on.

The particular method of finding clusters by alternately determining nearest neighbors of candidate centers and then locating neighborhood centers is also referred to as *K-means clustering*.

8. As presented here, the algorithm contains the idea's essence. In practice, to obtain good recognition, many refinements are necessary that are the result of intensive experimentation. For instance, the selection of initial cluster centers in step 3 may not in fact be carried out at random. This is the case with the vast majority of algorithms presented in this book: We describe the basic idea that must then be worked out in practice.

9. Five minutes of speech is usually adequate.

1.6 Additional Reading

The reader interested in finding out something about the human aspects of the speech production and perception process might start with Denes and Pinson [12] and continue with a more up-to-date article by Allen [13].

As pointed out in section 3.1, even though speech recognition depends crucially on the appropriate choice of signal processing, the latter is not a subject of this book. In addition to the references provided in that section [6] [7], it may be worthwhile to browse through a more recent book [14] or to consult articles by Cohen [15] and Picone [16] aimed specifically at signal processing for speech.

To simplify exposition of speech recognition's basic ideas and algorithms, we have limited ourselves to discrete outputs from the acoustic processor. The discretization is generally obtained by vector quantization (section 1.5) [11], about which it is possible to prove many interesting mathematical properties [17]. Unfortunately, vector quantization does in general lose some important information that would be available to the rest of the recognizer if it were fed by the signal processor's raw output.[10] More sophisticated methods of vector quantization can alleviate such loss [18] [19] [20]. Furthermore, a very interesting discretization method called *ranks* [21] both is robust and facilitates recognition results comparable to the very best in the state of the art. So dealing with discrete but sophisticated acoustic recognizer outputs turns out not to be much of a compromise after all.

In this book, we present what we think are our field's most fruitful approaches. Some leading contributors maintain that these are inadequate and that radical innovations are necessary to even approach the solution of the problem [22]. Twenty-eight years ago, a very famous communication engineer, N.R. Pierce, felt that ours was a hopeless endeavor [23].

References

[1] D.R. Reddy, "An approach to computer speech recognition by direct analysis of the speech wave," Tech. Report No. CS49, Computer Science Department, Stanford University, Sept. 1966.

[2] P. Ladefoged, *A Course in Phonetics*, Harcourt Brace Jovanovich, New York, 1975.

10. We do take up this eventuality in chapter 9.

[3] L.R. Bahl, F. Jelinek, and R.L. Mercer, "A maximum likelihood approach to continuous speech recognition," *IEEE Transactions on Pattern Analysis and Machine Intelligence*, vol. PAMI-5, pp. 179–90, March 1983.

[4] F. Jelinek, L.R. Bahl, and R.L. Mercer, "Design of a linguistic statistical decoder for the recognition of continuous speech," *IEEE Transactions on Information Theory*, vol. IT-21, pp. 250–56, 1975.

[5] A.F. Karr: *Probability*, Springer Verlag, New York, 1993.

[6] L.R. Rabiner and B-H Juang: *Fundamentals of Speech Recognition*, Prentice-Hall, Englewood Cliffs, NJ, 1993.

[7] L.R. Rabiner and R.W. Schafer, *Digital Processing of Speech Signals*, Prentice-Hall, Englewood Cliffs, NJ, 1978.

[8] R.P. Lippmann, "Review of neural networks for speech recognition," in *Readings in Speech Recognition*, A. Waibel and K.F. Lee, eds., Morgan Kaufmann, San Mateo, CA, 1990.

[9] L.R. Rabiner and S.E. Levinson, "Isolated and Connected Word Recognition—Theory and Selected Applications," *IEEE Transactions on Communications*, vol. COM-29, no. 5, pp. 621–69, May 1981.

[10] R.E. Ziemer and W.H. Tranter, *Principles of Communications*, John Wiley, New York, 1995.

[11] R.M. Gray, "Vector quantization," *IEEE ASSP Magazine*, vol. 1, pp. 4–29, April 1984.

[12] P.B. Denes and E.N. Pinson, *The Speech Chain: The Physics and Biology of Spoken Language*, Doubleday, New York, 1973.

[13] J.B. Allen, "How do humans process and recognize speech?" *IEEE Transactions on Speech and Audio Processing*, vol. 2, no. 4, pp. 567–77, October 1994.

[14] J.R. Deller, J.G. Proakis, and J.H.L. Hansen, *Discrete-Time Processing of Speech Signals*, Macmillan, New York, 1993.

[15] J.R. Cohen, "Application of an auditory model to speech recognition," *Journal of the Acoustic Society of America*, vol. 85, no. 6, pp. 2623–29, June 1989.

[16] J.W. Picone, "Signal modeling techniques in speech recognition," *Proceedings of the IEEE*, vol. 81, no. 9, pp. 1215–47.

[17] R.M. Gray and A. Gersho, *Vector Quantization and Signal Compression*, Klewer, Boston, MA, 1991.

[18] D. Rtischev, *Speaker Adaptation in a Large Vocabulary Speech Recognition System*, M.S. thesis, Massachussetts Institute of Technology, Cambridge, January 1989.

[19] A. Nadas, D. Nahamoo, and M.A. Picheny, "Speech recognition using noise adaptive prototypes," *IEEE Transactions on Acoustics, Speech, and Signal Processing*, vol. 37, no. 10, pp. 1495–1503, October 1989.

[20] L.R. Bahl, P.V. deSouza, P.S. Gopalakrishnan, and M.A. Picheny, "Context dependent vector quantization for continuous speech recognition," *Proceedings of*

the International Conference on Acoustics, Speech, and Signal Processing, vol. II, pp. 632–35, Minneapolis, MN, 1993.

[21] L.R. Bahl, P.V. deSouza, P.S. Gopalakrishnan, D. Nahamoo, and M.A. Picheny, "Robust methods for using context-dependent features and models in a continuous speech recognizer," *Proceedings of the International Conference on Acoustics, Speech, and Signal Processing*, vol. I, pp. 533–36, Adelaide, 1994.

[22] H. Bourlard, H. Hermansky, and N. Morgan, "Towards increasing speech recognition error rates," *Speech Communication*, vol. 18, no. 3, pp. 205–31, 1996.

[23] J.R. Pierce, "Whither speech recognition?" *Journal of the Acoustic Society of America*, vol. 46, pp. 1049–51, 1969.

Chapter 2

Hidden Markov Models

2.1 About Markov Chains

In chapter 1 we mentioned that the acoustic model that allows the recognizer to compute the probabilities $P(\mathbf{A}|\mathbf{W})$ is based on a concept referred to as a *hidden Markov model* (HMM).[1] This mathematical tool is very powerful and can be utilized for many purposes other than acoustic modeling. Before we apply it to the later (starting with chapter 3), we will discuss the HMM concept in general. But first some background.

A *Markov chain* is a very important and well-studied concept of probability theory. It deals with a class of random processes that incorporate a minimum amount of memory without actually being memoryless. Here we will deal with discrete random variables and finite Markov chains [2] [3].

Let $X_1, X_2, \ldots, X_n, \ldots$ be a sequence of random variables taking their values in the same finite alphabet $\mathcal{X} = \{1, 2, \ldots, c\}$. If nothing more is said, then Bayes' formula applies:

$$P(X_1, X_2, \ldots, X_n) = \prod_{i=1}^{n} P(X_i | X_1, X_2, \ldots, X_{i-1}) \tag{1}$$

The random variables are said to form a Markov chain, however, if

$$P(X_i | X_1, X_2, \ldots, X_{i-1}) = P(X_i | X_{i-1}) \qquad \text{for all values of } i \tag{2}$$

As a consequence, for Markov chains

$$P(X_1, X_2, \ldots, X_n) = \prod_{i=1}^{n} P(X_i | X_{i-1}) \tag{3}$$

---

1. In information theory, hidden Markov models are known as Markov sources. Shannon studied some of their aspects in his pioneering papers [1].

Such a random process thus has the simplest memory: The value at time i depends only on the value at the preceding time and on nothing that went on before.

The Markov chain is *time invariant* or *homogeneous* if regardless of the value of the time index i,

$$P(X_i = x'|X_{i-1} = x) = p(x'|x) \qquad \text{for all } x, x' \in \mathcal{X} \tag{4}$$

$p(x'|x)$ is called the *transition function* and it can be presented as a $c \times c$ matrix. Of course, $p(x'|x)$, which specifies everything about the random process, must for all $x \in \mathcal{X}$ satisfy the usual conditions

$$\sum_{x' \in \mathcal{X}} p(x'|x) = 1; \qquad p(x'|x) \geq 0 \quad x' \in \mathcal{X}$$

We can think of the values of X_i as *states* and thus of the Markov chain as a *finite state process* with transitions between states specified by the function $p(x'|x)$. If the alphabet $\mathcal{X}$ is not too large, then the chain can be completely specified by an intuitively appealing diagram such as figure 2.1, which applies to a state space of size $c = 3$. In the figure, arrows with attached transition probability values mark the transitions between states. Some transitions are missing, implying that $p(1|2) = p(2|2) = p(3|3) = 0$. The figure's shape shows why one might wish to call the process a *chain*.

The seeming restriction of Markov chains to one-step memory is deceiving. In principle, Markov chains are capable of modeling processes of arbitrary complexity. In fact, consider a process $Z_1, Z_2, \ldots, Z_n, \ldots$ of memory length k, that is,

$$P(Z_1, Z_2, \ldots, Z_n, \ldots) = \prod_i P(Z_i|Z_{i-k}, Z_{i-k+1}, \ldots, Z_{i-1})$$

If we define *new* random variables

$$X_i \doteq Z_{i-k+1}, Z_{i-k+2}, \ldots, Z_i \tag{5}$$

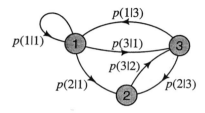

Figure 2.1
Three-state Markov chain

then the Z-sequence specifies the X-sequence, and vice versa, and moreover the X process is a Markov chain for which formula (3) holds. Of course, the resulting $\mathscr{X}$ state space is very large, and the Z process can be characterized directly in a much simpler way than via transformation (5).

2.2 The Hidden Markov Model Concept

We wish to allow more freedom to the random process while avoiding a substantial complication to the basic structure of Markov chains. We can gain this freedom by letting the states of the chain generate observable data while hiding the state sequence itself from the observer.

We thus define

1. An output alphabet $\mathscr{Y} = \{0, 1, \ldots, b-1\}$
2. A state space $\mathscr{S} = \{1, 2, \ldots, c\}$ with a unique starting state s_0.
3. A probability distribution of transitions between states $p(s'|s)$, and
4. An output probability distribution $q(y|s, s')$ associated with transitions from state s to state s'.

Then the probability of observing an HMM output string $y_1, y_2, \ldots, y_k$ is given by

$$P(y_1, y_2, \ldots, y_k) = \sum_{s_1, \ldots, s_k} \prod_{i=1}^{k} p(s_i|s_{i-1}) q(y_i|s_{i-1}, s_i) \tag{6}$$

Figure 2.2 is an example of an HMM with $b = 2$ and $c = 3$. There, we have attached the output distributions $q(y|s, s')$ to the transitions and have omitted the transition probabilities $p(s'|s)$.

Although the underlying state process still has only one-step memory,

$$P(s_1, s_2, \ldots, s_k) = \prod_{i=1}^{k} p(s_i|s_{i-1})$$

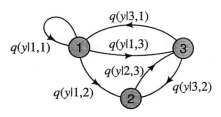

Figure 2.2
Three-state hidden Markov model with outputs $y \in \{0, 1\}$

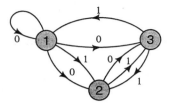

Figure 2.3
Hidden Markov model representation attaching outputs to transitions

the memory of observables is unlimited (except in degenerate cases). That is, in general, for all $k \geq 2$,

$$P(y_{k+1}|y_1, y_2, \ldots, y_k) \neq P(y_{k+1}|y_j, \ldots, y_k) \qquad k \geq j \geq 2$$

We will frequently find it convenient to regard the HMM as having multiple transitions between pairs of states, each associated with a different output symbol generated, with probability 1, when that transition is taken. This is directly equivalent to the formulation of the present section. Figure 2.3 gives an example that can generate the same random process as figure 2.2.[2]

This view has the advantage of allowing us to provide each transition of the entire HMM with a different identifier t and to define an output function $Y(t)$ that assigns to t a unique output symbol taken from the alphabet $\mathcal{Y}$.

We then denote by $L(t)$ and $R(t)$ the source and target states of the transition t, respectively. We let $p(t)$ denote the probability that the state $L(t)$ is exited via the transition t, so that for all $s \in \mathcal{S}$,

$$\sum_{t:L(t)=s} p(t) = 1$$

The correspondence between the two ways of viewing an HMM is given by the relationship

$$p(t) = q(Y(t)|L(t), R(t))p(R(t)|L(t)) \qquad (7)$$

When transitions determine outputs, the probability $P(y_1, y_2, \ldots, y_k)$ becomes equal to the sum of the products $\prod_{i=1}^{k} p(t_i)$ over all transition sequences $t_1, \ldots, t_k$ such that $L(t_1) = s_0$, $Y(t_i) = y_i$, and $R(t_i) = L(t_{i+1})$ for $i = 1, \ldots, k$, or, formally,

---

2. Assuming that in figure 2.2 $q(1|1, 1) = q(1|1, 3) = q(0|3, 1) = q(0|3, 2) = 0$

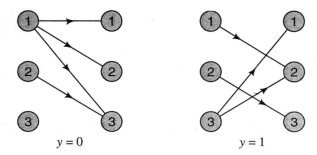

Figure 2.4
Two different trellis stages corresponding to the binary HMM of figure 2.3

$$P(y_1, y_2, \ldots, y_k) = \sum_{\mathscr{S}(y_1, y_2, \ldots, y_k)} \prod_{i=1}^{k} p(t_i)$$

where

$$\mathscr{S}(y_1, y_2, \ldots, y_k) = \{t_1, \ldots, t_k : L(t_1) = s_0, Y(t_i) = y_i, R(t_i) = L(t_{i+1})$$
$$\text{for } i = 1, \ldots, k\}$$

In the following sections we will take whichever point of view (multiple transitions between states s and s', or multiple possible outputs generated by the single transition $s \rightarrow s'$) will be more convenient for the problem at hand.

2.3   The Trellis

There is an easy way to calculate the probability $P(y_1, y_2, \ldots, y_k)$ with the help of a *trellis*. The trellis places in evidence the time evolution of the process that generated the sequence $y_1, y_2, \ldots, y_k$. It consists of the concatenation of elementary *stages* determined by the particular outputs y_i. There are as many different elementary stages as there are different output symbols. Figure 2.4 shows the stages for $y_i = 0$ and $y_i = 1$ corresponding to the HMM of figure 2.3. Note that a stage for y_i consists of left and right columns of c states connected by those transitions t whose output $Y(t) = y_i$. Figure 2.5 shows the trellis corresponding to the output sequence 0110. The required probability $P(0110)$ is equal to the sum of the probabilities of all complete paths through the trellis (those ending in the last column) that start in the obligatory starting state. Figure 2.6 shows these paths for $s_0 = 1$.

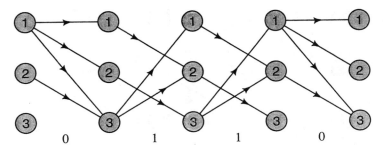

Figure 2.5
Trellis for output sequence 0110 generated by the HMM of figure 2.3

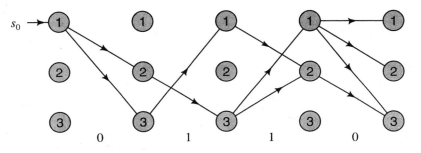

Figure 2.6
Trellis of figure 2.5 purged of all paths that could not have generated the output sequence 0110

The probability $P(y_1, y_2, \ldots, y_n)$ can be obtained recursively. Define the probabilities

$$\alpha_i(s) \doteq P(y_1, y_2, \ldots, y_i, s_i = s) \qquad (8)$$

Then with boundary conditions

$$\alpha_0(s) = 1 \text{ for } s = s_0 \text{ and } \alpha_0(s) = 0 \text{ otherwise}$$

we get the recursion

$$\alpha_i(s) = \sum_{s'} p(y_i, s | s') \alpha_{i-1}(s') \qquad (9)$$

where we use the simplified notation

$$p(y_i, s | s') = q(y_i | s', s) p(s | s') \qquad (10)$$

By definition (8) the desired probability then is

$$P(y_1, y_2, \ldots, y_k) = \sum_s \alpha_k(s) \tag{11}$$

The recursion (9) for the output sequence 0110 is easily visualized on the trellis of figure 2.5. Unit probability is assigned to the starting state 1 in the 0th column. The values $\alpha_1(s)$ are then attached to the states of column 1 by computing the flows $p(0, s|1)\alpha_0(1)$ for $s \in \{1, 2, 3\}$. Then the values $\alpha_2(s)$ are attached to the states of column 2 by summing the flows $p(1, s|s')\alpha_1(s')$ that leave the various states s' of column 1, etc. Finally, the probability $P(0110)$ is obtained by summing the values $\alpha_4(s)$ attached to the states of the fourth column.

2.4 Search for the Likeliest State Transition Sequence

Given an observed output sequence $y_1, y_2, \ldots, y_k$, which state sequence $s_1, s_2, \ldots, s_k$ is most likely to have caused it? Noting that

$$P(s_1, s_2, \ldots, s_k | y_1, y_2, \ldots, y_k, s_0) = \frac{P(s_1, s_2, \ldots, s_k, y_1, y_2, \ldots, y_k | s_0)}{P(y_1, y_2, \ldots, y_k | s_0)}$$

we see that we want to find the sequence $s_1, s_2, \ldots, s_k$ maximizing the numerator of the right-hand side of the above equation.

Observe that because the underlying state process is Markov, then for all i,

$$P(s_1, \ldots, s_i, s_{i+1}, \ldots, s_k, y_1, y_2, \ldots, y_i, y_{i+1}, \ldots, y_k | s_0)$$

$$= P(s_1, s_2, \ldots s_i, y_1, y_2, \ldots, y_i | s_0) P(s_{i+1}, \ldots, s_k, y_{i+1}, \ldots, y_k | s_i)$$

so that

$$\max_{s_1, s_2, \ldots, s_k} P(s_1, \ldots, s_i, s_{i+1}, \ldots, s_k, y_1, y_2, \ldots, y_i, y_{i+1}, \ldots, y_k | s_0)$$

$$= \max_{s_i, s_{i+1}, \ldots, s_k} \left[P(s_{i+1}, \ldots, s_k, y_{i+1}, \ldots, y_k | s_i) \right. \tag{12}$$

$$\left. \times \max_{s_1, s_2, \ldots, s_{i-1}} P(s_1, s_2, \ldots, s_{i-1}, s_i, y_1, y_2, \ldots, y_i | s_0) \right]$$

Defining

$$\gamma_i(s_i) \doteq \max_{s_1, s_2, \ldots, s_{i-1}} P(s_1, s_2, \ldots s_i, y_1, y_2, \ldots, y_i | s_0) \tag{13}$$

and substituting into (12), we get

$$\max_{s_1, s_2, \ldots, s_k} P(s_1, \ldots, s_i, s_{i+1}, \ldots, s_k, y_1, y_2, \ldots, y_i, y_{i+1}, \ldots, y_k | s_0)$$

$$= \max_s \left\{ \max_{s_{i+1}, \ldots, s_k} [P(s_{i+1}, \ldots, s_k, y_{i+1}, \ldots, y_k | s) \gamma_i(s)] \right\} \qquad (14)$$

Therefore, we can find the maximally likely sequence $s_1, \ldots, s_{i-1}, s_i$, $s_{i+1}, \ldots, s_k$ by finding first for each state s on trellis level i the most likely state sequence $s_1(s), \ldots, s_{i-1}(s)$ leading into s, then finding the most likely sequence $s_{i+1}(s), \ldots, s_k(s)$ leading out of s, and finally finding the state s on level i for which the complete sequence $s_1(s), \ldots, s_{i-1}(s), s_i = s, s_{i+1}(s)$, $\ldots, s_k(s)$ has the highest probability.

This can be done recursively. In fact, it follows directly from definition (13) and the HMM defining rule (10) that

$$\gamma_i(s_i) = \max_{s_1, \ldots, s_{i-1}} P(s_1, s_2, \ldots, s_i, y_1, y_2, \ldots, y_i | s_0)$$

$$= \max_{s_{i-1}} p(y_i, s_i | s_{i-1}) \max_{s_1, \ldots, s_{i-2}} P(s_1, s_2, \ldots, s_{i-1}, y_1, y_2, \ldots, y_i | s_0) \qquad (15)$$

$$= \max_{s_{i-1}} p(y_i s_i | s_{i-1}) \gamma_{i-1}(s_{i-1})$$

Noticing from (13) that

$$\max_{s_1, \ldots, s_k} P(s_1, s_2, \ldots, s_k, y_1, y_2, \ldots, y_k | s_0) = \max_s \gamma_k(s)$$

equation (15) leads directly to the *Viterbi algorithm* [4] that finds a maximizing state sequence (there may be more than one) by finding the maximizing sequence $s_1(s), \ldots, s_{i-1}(s), s_i = s$ (whose probability is $\gamma_i(s)$) for each state s on successive levels i, and finally deciding at level k from among the competing sequences $s_1(s), \ldots, s_{k-1}(s), s_k = s, s \in \mathcal{S}$.

Here then is a complete statement of the Viterbi algorithm.

The Viterbi Algorithm

1. Set $\gamma_0(s_0) = 1$.
2. Use (15) to compute $\gamma_1(s)$ for all states s of the trellis's first column, that is,

$$\gamma_1(s) = \max_{s'} p(y_1, s | s') \gamma_0(s') = p(y_1, s | s_0)$$

since $\gamma_0(s') = 0$ for all $s' \neq s_0$.

3. Compute $\gamma_2(s)$ for all states s of the trellis's second column, that is,

$$\gamma_2(s) = \max_{s'} p(y_2, s | s') \gamma_1(s')$$

Purge all transitions from states s' in the first column to states s in the

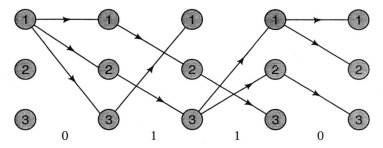

Figure 2.7
Possible result of applying the Viterbi algorithm to the trellis of figure 2.5

second column for which $\gamma_2(s) > p(y_2, s|s')\gamma_1(s')$. If more than one transition into state s remains, select (arbitrarily) one to keep and purge the rest.
4. In general, compute $\gamma_i(s)$ for all states s of the trellis's i^{th} column. Purge all transitions from states s' in the $i - 1^{th}$ column to states s in the ith column for which $\gamma_i(s) > p(y_i, s|s')\gamma_{i-1}(s')$. Then purge all but one of the remaining transitions into s.
5. Find the state s in the trellis's k^{th} column for which $\gamma_k(s)$ is maximal. In the purged trellis, trace back from this state s to the initial state s_0 in the 0th column along the remaining transitions. The states $s_1, s_2, \ldots, s_k = s$ encountered along this path constitute a most likely state sequence.

Figure 2.7 shows an example of a possible effect of the Viterbi algorithm on the trellis of figure 2.5 that corresponds to the observed sequence 0110. Assuming that the largest $\gamma_4(s)$ is achieved for $s = 3$, the traceback gives as a solution the state sequence 12323.

2.5 Presence of Null Transitions

In many applications, and in speech recognition in particular, it is useful to introduce the possibility of *null* transitions that change state but result in no output. In figure 2.8 we have added null transitions (interrupted lines) to the HMM of figure 2.3.

Since a trellis stage corresponds to a single output, null transitions take place between the states in the initial column of a stage. Figure 2.9 is then the trellis corresponding to the output 0110.

We must now make the appropriate changes in our formulas to accommodate the presence of null transitions. The changes made in this section are completely general and result in defining an equivalent HMM that

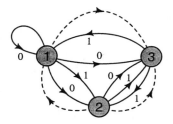

Figure 2.8
Addition of null transitions to the HMM of figure 2.3

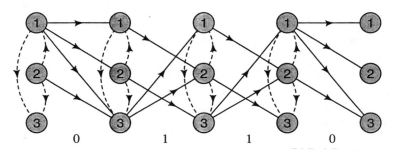

Figure 2.9
Trellis corresponding to the output 0110 generated by the HMM of figure 2.8

has no null transitions. In the next section we address the special case when null transitions do not form a loop.

First, the recursion (9) for $\alpha_i(s)$ is still valid, provided $p(y_i, s|s')$ is properly defined as the probability of starting in state s' in trellis column $i - 1$, making any number of null transitions and then taking an output generating transition t that produces y and has a final state s.

Let $q(s'|s)$ be the probability of a null transition from s to s'. Let $r(y, s'|s)$ be the probability of a (non-null) transition from s to s' producing the output y. We must have

$$\sum_{y,s'} r(y, s'|s) + \sum_{s''} q(s''|s) = 1$$

Define a new matrix

$$v(s'|s) \doteq \sum_{i=0}^{\infty} q^{(i)}(s'|s) \tag{16}$$

where

$$q^{(0)}(s'|s) \doteq \begin{cases} 1 & \text{if } s' = s \\ 0 & \text{otherwise} \end{cases}$$

and $q^{(i)}(s'|s)$ denotes the probability of reaching s' from s in i null transitions. This probability is, of course, the (s, s') element of the ith product of the matrix $\mathbf{q}(s'|s)$ with itself, or equivalently,

$$q^{(i)}(s'|s) = \sum_{s''} q^{(i-1)}(s''|s)q(s'|s'')$$

The appropriate definition then is

$$p(y, s|s') \doteq \sum_{s''} r(y, s|s'')v(s''|s') \tag{17}$$

Note that if there is no loop of null transitions then necessarily $q^{(i)}(s'|s) = 0$ for $i \geq c$, where c is the number of states.[3]

Again, the probability $p(y, s|s')$ defined by (17) is simply the probability that starting from state s', the next non-null transition taken is t where $R(t) = s$ and $Y(t) = y$.[4] It is easy to check that $p(y, s|s')$ as defined above is a proper probability in that

$$\sum_{y,s} p(y, s|s') = 1$$

Thus $p(y, s|s')$ defines a new HMM that has no null transitions and generates output strings with exactly the same probability as does the original HMM with parameters $r(y, s|s')$ and $q(s'|s)$. Figure 2.10 shows the structure of this type of HMM that is equivalent to the HMM of figure 2.8.

2.6 Dealing with an HMM That Has Null Transitions That Do Not Form a Loop

If it is desired to find the most likely state sequence $s_1, s_2, \ldots, s_m$ of the original HMM given an observed output sequence $y_1, y_2, \ldots, y_k$, we must

3. In any case, it is easy to show from definition (16) that

$$[\mathbf{v}(s'|s)] = [\mathbf{I} - \mathbf{q}(s'|s)]^{-1}$$

where $\mathbf{I}$ denotes the identity matrix. The inverse exists as long as $\sum_{s'} q(s'|s) < 1$ for all s.

4. This path consists of any number (including 0) of null transitions followed by the output producing transition t.

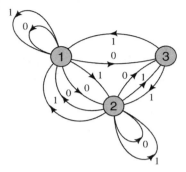

Figure 2.10
HMM without null transitions equivalent to the HMM of figure 2.8

proceed differently. Note that now $m \geq k$. First we need a new definition

$$\gamma_i(s) \doteq \max_{s_1, \ldots, s_{l-1}} P(s_1, s_2, \ldots, s_{l-1}, s_l = s, y_1, y_2, \ldots, y_i \mid s_0)$$

where it is understood that the state sequence $s_1, s_2, \ldots s_{l-1}, s_l$ ends in the trellis's i^{th} column.[5] The desired recursion then becomes

$$\gamma_i(s) = \max \left\{ \max_{s'} r(y_i, s \mid s') \gamma_{i-1}(s'), \max_{s'' < s} q(s \mid s'') \gamma_i(s'') \right\} \tag{18}$$

But note from (18) that to evaluate $\gamma_i(s)$ we need to know $\gamma_i(s'')$ for all states from which a null transition into s is possible (this is what is meant by the condition $s'' < s$). Thus it must be possible to order the states appropriately so that (18) can be evaluated in sequence.

Such ordering is indeed possible provided the null transitions do not form a loop in the HMM of interest. The following is a simple ordering algorithm:

1. Place into set $\mathscr{S}_1$ all states that cannot be reached from any other state via a null transition. (If null transitions do not form a loop, there must be at least one state with no null transitions leading into it, so $\mathscr{S}_1$ is not empty.)
2. Place into set $\mathscr{S}_2$ all states that can be reached via a single null transition only from states of set $\mathscr{S}_1$.

5. A really careful definition might be something like

$$\gamma_i(s) \doteq \max_{s_1, \ldots, s_{l-1}} P(s_1, \ldots, s_{l-1}, s_l = s, \mathbf{Y}(s_0, s_1, \ldots, s_{l-1}, s_l) = y_1 \cdots y_{i-1} y_i \mid s_0)$$

where $\mathbf{Y}(s_0, s_1, \ldots, s_{l-1}, s_l)$ denotes the total output sequence resulting from the state sequence $s_0 \to s_1 \to \cdots \to s_{l-1} \to s_l$.

3. In general, place into set $\mathscr{S}_i$ all states that can be reached via a single null transition only from states in $\mathscr{S}_{i-j}, j = 1, 2, \ldots, i - 1$.

4. Stop when no states remain to be placed. Index states in a sequence such that states in $\mathscr{S}_{i-1}$ have a lower index than those in $\mathscr{S}_i$ (i.e., assign the lowest indices to states in $\mathscr{S}_1$, the next indices to states in $\mathscr{S}_2$, etc.).

In the HMM of figure 2.8, the set $\mathscr{S}_1$ contains state 2, the set $\mathscr{S}_2$ contains state 1, and set $\mathscr{S}_3$ contains state 3.

For HMMs whose null transitions do not form a loop, just as we adjusted (15) into (17), we can similarly adjust formulas (8) through (11):

$$\alpha_i(s) \doteq P(y_1, y_2, \ldots, y_i, s^i = s) \tag{19}$$

where s^i is to be interpreted as a state in the ith trellis column (i.e., not as the ith state reached, which we denote by s_i). With this interpretation we get the recursion

$$\alpha_i(s) = \sum_{s'} r(y_i, s|s')\alpha_{i-1}(s') + \sum_{s''<s} q(s|s'')\alpha_i(s'') \tag{20}$$

2.7 Estimation of Statistical Parameters of HMMs

So far we have acted as if some higher authority—or the problem itself—specifies to us the parameters (transition and output production probabilities) of HMMs. This is, of course, rarely the case. We want to use HMMs to model data generation, and the most we can expect is to be given a sample of such data. We then want to construct a model that will account for future (as yet unseen) data whose type is the same as that of the given sample.[6]

It turns out that there exists no good way to estimate both the *transition structure* and the statistical parameters of an HMM. The best we can do is to use our knowledge of the situation and our intuition to design the HMM structure, and from that, estimate the parameter values. We would like to have a method that has the following *maximum likelihood* property:

Let $P_\lambda(\mathbf{Y})$ denote the probability that the HMM defined by parameters λ produces the observed output $\mathbf{Y} = y_1, y_2, \ldots, y_k$. We wish to find

$$\hat{\lambda} = \arg \max_\lambda P_\lambda(\mathbf{Y}) \tag{21}$$

---

6. We will sidestep a possible philosophical argument as to what *same type* might mean and rely on our intuition.

The parameter specification $\hat{\lambda}$ would, in an important statistical sense, allow the HMM (of the given structure) to account best for the observed data **Y**. We should bear in mind, however, that what we would really like is for $\hat{\lambda}$ to account for as yet unseen data that will come to us later for processing. We cannot count on this. We can only entertain the intuitively attractive hope that the *training data* **Y** is sufficiently representative of the process so that $\hat{\lambda}$ will remain satisfactory (but not necessarily optimal) for unseen data as well.

We will arrive at the required algorithm (known in the literature variously as the Baum, Baum-Welch, or forward-backward algorithm) by intuitive reasoning. Proofs of its convergence, optimality, or other characteristics can be found in many references (e.g., [5] [6]) and will be omitted in this chapter. In chapter 9, however, we will show one proof based on the *expectation-maximization* algorithm.

The following heuristic reasoning will seemingly imply that the data **Y** was actually produced by an HMM that has the designed structure. The reader should not be fooled by this pedagogical device: The machinery that produced **Y** remains unknown to us.

It will be convenient to use the following HMM representation in this section: Between any pair of states s and s' there exists at most (a) one output producing transition t with $L(t) = s$ and $R(t) = s'$ and (b) one null transition t' with $L(t') = s$ and $R(t') = s'$. We will also use the simplified notation

$$q(y|t) \doteq q(y|s, s')$$

It is best to gain the appropriate feeling for the task of estimating HMM parameters by means of a simple example that involves the basic phenomenon we wish to treat. We will consider the HMM of figure 2.11

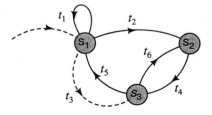

Figure 2.11
Example HMM with indexed transitions producing binary outputs

that produces binary sequences. There are three states and six transitions, t_3 being a null transition. The transition probabilities satisfy constraints

$$p(t_1) + p(t_2) + p(t_3) = 1; \quad p(t_5) + p(t_6) = 1$$

We will wish to estimate the probabilities $p(t_i)$ and $q(y|t_i)$ where $y \in \{0, 1\}$ and $i \in \{1, \ldots, 6\}$. We will designate $s_0 = 1$ as the starting state.

Suppose we knew that the HMM produced the (long) sequence $y_1, \ldots, y_k$. How would we estimate the required parameters? A good way to see the possible operation of the HMM is to develop it in time by means of a trellis, such as that of figure 2.12, corresponding to the production of y_1, y_2, y_3, y_4 by the HMM of figure 2.11. In figure 2.12 we have added superscripts to transitions and states to indicate the trellis stage from which the transitions originate or in which the states are located. The output sequence can be produced by any path leading from $s_0 = 1$ in the 0th stage to any of the states in the final stage.

Suppose we knew the actual transition sequence $\mathbf{T} = t_1, t_2, \ldots, t_m$ ($m \geq k$, because some transitions may be null transitions and k outputs were generated) that caused the observed output. Then the so called *maximum likelihood* parameter estimate could be obtained by counting. Define the indicator function,

$$I_i(t) \doteq \begin{cases} 1 & \text{if one of the transitions in } \mathbf{T} \text{ from a state} \\ & \text{belonging to } i^{th} \text{ stage is } t \\ 0 & \text{otherwise} \end{cases} \tag{22}$$

and using it,

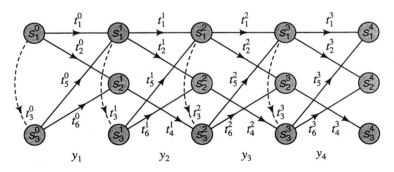

Figure 2.12
Trellis for the HMM of figure 2.11 corresponding to the output sequence y_1, y_2, y_3, y_4

$$c(t) \doteq \sum_{i=0}^{k-1} I_i(t) \tag{23}$$

for all transitions t, as well as

$$c(y, t') \doteq \sum_{i=0}^{k-1} I_i(t') \delta(y_{i+1}, y) \tag{24}$$

for non-null transitions t'. In (24), $\delta(\ ,\ )$ denotes the Kronecker delta function.[7] The number of times the process went through transition t is equal to $c(t)$, and $c(y, t)$ is the number of times the process went through t and as a consequence produced the output y.

The "natural" estimates then would be

$$\hat{q}(y|t) = \frac{c(y, t)}{c(t)} \qquad t \text{ non-null} \tag{25}$$

$$\hat{p}(t_i) = \begin{cases} \dfrac{c(t_i)}{c(t_1) + c(t_2) + c(t_3)} & i = 1, 2, 3 \\[2mm] 1 & i = 4 \\[2mm] \dfrac{c(t_i)}{c(t_5) + c(t_6)} & i = 5, 6 \end{cases} \tag{26}$$

Of course, the idea of our knowing the transition sequence $t_1, t_2, \ldots, t_m$ is absurd. But what if, knowing that $y_1, y_2, \ldots, y_k$ was observed, we calculated the probabilities $P\{t^i = t\}$ that transition t was taken when leaving the ith stage of the process?[8] Then it would seem intuitively reasonable to define the "counts" by (t' is restricted to non-null transitions)

$$c(t) \doteq \sum_{i=0}^{k-1} P\{t^i = t\} \tag{27}$$

$$c(y, t') \doteq \sum_{i=0}^{k-1} P\{t^i = t'\} \delta(y_{i+1}, y) \tag{28}$$

and use these in the estimation formulas (25) and (26).

7. The definition is
$$\delta(y, y') = \begin{cases} 1 & \text{if } y = y' \\ 0 & \text{otherwise} \end{cases}$$
8. This is not necessarily the ith transition. It is the ith output producing transition.

Because the production of each output y_i corresponds to some non-null transition from stage $i - 1$ to stage i, then $\sum P\{t^i = t''\} = 1$, where the sum is over all non-null transitions t''. Thus whereas in formula (23), the counter of exactly one of the non-null transitions gets increased by 1 for each output, this same contribution is simply distributed by (27) among the various counters belonging to non-null transitions.

Actually, it will be more convenient to use the renormalized count estimates

$$\hat{q}(y|t) = \frac{c^*(y,t)}{c^*(t)} \qquad t \text{ non-null}$$

$$\hat{p}(t) = \frac{c^*(t)}{\sum_{t':L(t')=L(t)} c^*(t')}$$

where $c^*(y,t) \doteq c(y,t)P(y_1, y_2, \ldots, y_k)$ and $c^*(t) \doteq c(t)P(y_1, y_2, \ldots, y_k)$.[9]

The estimates arising from (27) and (28) will be convenient only if it proves to be practical to compute the required probabilities

$$P^*\{t^i = t\} \doteq P\{t^i = t\}P(y_1, y_2, \ldots, y_k) \tag{29}$$

To derive the appropriate formulas, we need some precise notation in addition to the one already defined:

$P^*\{t^i = t\}$ the probability that $y_1, \ldots, y_k$ was produced and the transition out of a state in the ith trellis stage was t; $i = 0, 1, \ldots, k - 1$

$\alpha_i(s) = P\{s^i = s\}$ the probability that $y_1, y_2, \ldots, y_i$ was produced and the state reached at the ith stage was s; $i = 0, 1, \ldots, k$

$\beta_i(s) = P\{rest|s^i = s\}$ the probability that $y_{i+1}, \ldots, y_k$ was produced when the state at the ith stage was s; $i = 0, 1, \ldots, k$ (30)

Since the transition t can be taken only after the HMM has reached the state $L(t)$, and since after it is taken, the rest of the action will start in state $R(t)$, we see immediately that

---

9. Note that $P\{t^i = t\}$ is defined as a probability conditioned on the observed sequence $y_1, \ldots, y_k$. Therefore, $P\{t^i = t\}P(y_1, \ldots, y_k)$ is the corresponding joint probability. The point, as we shall see, is that this joint probability is directly computable.

$$P^*\{t^i = t\} = \begin{cases} \alpha_i(L(t))\,p(t)\,q(y_{i+1}|t)\,\beta_{i+1}(R(t)) & \text{if } t \text{ is non-null} \\ \alpha_i(L(t))\,p(t)\,\beta_i(R(t)) & \text{if } t \text{ is null} \end{cases} \tag{31}$$

It is also obvious that the following recursion holds (it is identical to (20)):

$$\alpha_i(s) = \sum_{t \in \bar{\mathcal{N}}(s)} \alpha_{i-1}(L(t))\,p(t)\,q(y_i|t) + \sum_{t \in \mathcal{N}(s)} \alpha_i(L(t))\,p(t) \tag{32}$$

Here $\mathcal{N}(s)$ is the set of all null transitions from states of lower order than s and ending in s and $\bar{\mathcal{N}}(s)$ is the set of all non-null transitions ending in s. That is,

$$\mathcal{N}(s) = \{t : R(t) = s, t \text{ is null}, L(t) < s\}$$
$$\bar{\mathcal{N}}(s) = \{t : R(t) = s, t \text{ is non-null}\} \tag{33}$$

Even though (32) involves quantities $\alpha_i(s)$ on both sides of the equation, it can be easily evaluated whenever null transitions do not form a loop (as in the example HMM), because in such a case the states can be appropriately ordered by the algorithm we introduced in section 2.6.

In addition to the *forward* recursion (32), we can also obtain a *backward* recursion:

$$\beta_i(s) = \sum_{t \in \bar{\mathcal{M}}(s)} p(t)\,q(y_{i+1}|t)\,\beta_{i+1}(R(t)) + \sum_{t \in \mathcal{M}(s)} p(t)\,\beta_i(R(t)) \tag{34}$$

where $\mathcal{M}(s)$ and $\bar{\mathcal{M}}(s)$ are appropriate sets of null and non-null transitions leaving s, respectively:

$$\mathcal{M}(s) = \{t : L(t) = s, t \text{ is null}, R(t) > s\}$$
$$\bar{\mathcal{M}}(s) = \{t : L(t) = s, t \text{ is non-null}\} \tag{35}$$

Recursions (32) and (34) then lead directly to an algorithm computing the desired quantities $P^*\{t^i = t\}$ via formula (31):

1. *The forward pass:* Setting $\alpha_0(s_0) = 1$ (in all of our example HMMs, $s_0 = 1$), use (32) to compute $\alpha_i(s)$ for all s, starting with $i = 0$ and ending with $i = k$.
2. *The backward pass:* Setting $\beta_k(s) = 1$ for all s, use (34) to compute $\beta_i(s)$ for all s, starting with $i = k - 1$ and ending with $i = 0$.
3. *Transition probability evaluation:* Using formula (31), evaluate probabilities $P^*\{t^i = t\}$ for all t and $i = 0, 1, \ldots, k - 1$.

4. *Parameter estimation:* Use the "counts"

$$c^*(t) \doteq \sum_{i=0}^{k-1} P^*\{t^i = t\}$$

and

$$c^*(y, t') \doteq \sum_{i=0}^{k-1} P^*\{t^i = t'\} \delta(y_{i+1}, y)$$

to get the parameter estimates

$$\hat{q}(y|t) = \frac{c^*(y, t)}{c^*(t)}, \quad t \text{ non-null} \tag{36}$$

$$\hat{p}(t) = \frac{c^*(t)}{\sum_{t':L(t')=L(t)} c^*(t')} \tag{37}$$

There is only one flaw, seemingly a fatal one, to our procedure: Formulas (31), (32), and (34) use the values of the parameters $p(t)$ and $q(y|t)$ that we are trying to estimate! Fortunately, there is a good way out: We put the above algorithm into a loop, start with a guess at $p(t)$ and $q(y|t)$, obtain better estimates (this is proved in chapter 9) for (36) and (37), then plug these back in for $p(t)$ and $q(y|t)$ to run the algorithm again, obtaining an even better estimate, etc., etc.

We cannot guarantee, however, that the Baum algorithm just described converges to the desired parameter set $\hat{\lambda}$ defined in (21). Instead, the algorithm finds only a local maximum. More precisely,

Let $P_\lambda(\mathbf{Y})$ denote the probability that the HMM defined by parameters λ produced the observed output $\mathbf{Y}$. If λ' denotes the parameter values determined by (36) and (37) when parameters λ were used in (31), (32), and (34), then $P_{\lambda'}(\mathbf{Y}) \geq P_\lambda(\mathbf{Y})$

2.8 Practical Need for Normalization

Often, the recursions (32) and (34) need to be carried out for a large number of stages k. As a result, the values of $\alpha_i(s)$ (or $\beta_i(s)$) lose precision as i increases (decreases). To prevent this loss from seriously affecting the estimates of the HMM parameter values, it is necessary to carry out some sort of normalization as the recursion progresses. One good way is as follows:

Let $Q_0, Q_1, Q_2, \ldots, Q_k$ be positive numbers, and define

$$\alpha_i^*(s) \doteq \frac{\alpha_i(s)}{\prod_{j=0}^{i} Q_j} \tag{38}$$

Then the recursion (32) becomes

$$\alpha_i'(s) \doteq \alpha_i^*(s) Q_i = \sum_{t \in \mathcal{N}(s)} \alpha_{i-1}^*(L(t)) \, p(t) \, q(y_i | t) + \sum_{t \in \bar{\mathcal{N}}(s)} \alpha_i'(L(t)) \, p(t) \tag{39}$$

where the sets $\mathcal{N}(s)$ and $\bar{\mathcal{N}}(s)$ are defined in (33).

If we then chose

$$Q_0 = \sum_s \alpha_0(s), \qquad Q_i = \sum_s \alpha_i'(s), i = 0, 1, 2, \ldots, k \tag{40}$$

we will assure that $\sum_s \alpha_i^*(s) = 1$ for all values of i, and that the forward recursion process is properly normalized.[10] Note that for $i = 0$, the first righthand side of (39) is missing, $\alpha_0'(s) = \alpha_0(s)$, $\alpha_0(s) = 0$ for $s < s_0$, and $\alpha_0(s_0) = 1$.

It is very convenient to use the same factors Q_j to normalize the backward recursion:

$$\beta_i^*(s) \doteq \frac{\beta_i(s)}{\prod_{j=i}^{k} Q_j} \tag{41}$$

Then similarly

$$\beta_i'(s) \doteq \beta_i^*(s) Q_i = \sum_{t \in \mathcal{M}(s)} p(t) \, q(y_{i+1} | t) \, \beta_{i+1}^*(R(t)) + \sum_{t \in \bar{\mathcal{M}}(s)} p(t) \, \beta_i'(R(t))$$

where the sets $\mathcal{M}(s)$ and $\bar{\mathcal{M}}(s)$ are defined in (35).

Note next that as a result of the choice (40),

$$P(y_1, y_2, \ldots, y_k) = \sum_s \alpha_k(s) = \sum_s \left[\alpha_k^*(s) \prod_{i=0}^{k} Q_i \right] = \prod_{i=0}^{k} Q_i$$

because by definition of the normalization process, $\sum_s \alpha_k^*(s) = 1$. Then it follows from (38) and (41) that for all s, s', and $i = 0, 1, \ldots, k - 1$,

10. A stage of the forward process then is as follows:

1. Using normalized values $\alpha_{i-1}^*(s')$ in (39), obtain values $\alpha_i'(s)$ in the natural order of the states s.
2. Calculate $Q_i = \sum_s \alpha_i'(s)$.
3. By normalization $\alpha_i^*(s) = \alpha_i'(s)/Q_i$ obtain the α-values necessary for calculating $\alpha_{i+1}'(s)$.

$$\alpha_i(s)\beta_{i+1}(s') = \alpha_i^*(s)\beta_{i+1}^*(s') \times \prod_{j=0}^{k} Q_j$$

$$= \alpha_i^*(s)\beta_{i+1}^*(s') \times P(y_1, y_2, \ldots, y_k)$$

As a consequence of definition (29) and formula (31), we may then use the original conditional probabilities

$$P\{t^i = t\} = \begin{cases} \alpha_i^*(L(t))\, p(t)\, q(y_{i+1}|t)\, \beta_{i+1}^*(R(t)) & \text{if } t \text{ non-null} \\ \alpha_i^*(L(t))\, p(t)\, \beta_i^*(R(t))\, Q_i & \text{if } t \text{ null} \end{cases} \tag{42}$$

as contributions to the estimation counters $c(y, t)$ and $c(t)$ (see (27) and (28)) thus obviating the calculation of the product $\prod_{j=0}^{k} Q_j$.

2.9 Alternative Definitions of HMMs

2.9.1 HMMs Outputting Real Numbers

The HMM concept can be generalized in a way that is very useful to acoustic modeling. The finite size restriction on the output alphabet $\mathcal{Y}$ can be removed. Thus outputs y can consist of real numbers or even real vectors. In that case the output probabilities become densities and their estimation must be parametric. For instance, a Gaussian assumption about the density $q(y|t)$ is made, and the Baum algorithm estimates its means and variances. We will carry out this generalization in chapter 9, which deals with the Expectation-Maximization algorithm that can be used to prove rigorously the convergence of the Baum algorithm.

2.9.2 HMM Outputs Attached to States

Many researchers prefer dealing with a structure in which outputs are generated by states and not by transitions into states. Thus their HMM is defined by

1. An output alphabet $\mathcal{Y} = \{0, 1, \ldots, b - 1\}$
2. A state space $\mathcal{S} = \{1, 2, \ldots, c\}$ with a unique starting state s_0
3. A probability distribution of transitions between states $p(s'|s)$, and
4. An output probability distribution (or density) $q(y|s)$

Then the probability of observing an HMM output string $y_1, y_2, \ldots, y_k$ is given by

$$P(y_1, y_2, \ldots, y_k) = \sum_{s_1, \ldots, s_k} \prod_{i=1}^{k} p(s_i|s_{i-1}) q(y_i|s_i) \tag{43}$$

The two HMM formulations (this and that of section 2.2) are entirely equivalent. In fact,

1. To get the setup of (43) from that of (6)[11], we enlarge the state space of the latter to be of size c^2 with states $\sigma = (s, s')$, and define an output function $q^*(y|\sigma) = q(y|s, s')$ and a state transition function

$$p^*(\sigma''|\sigma) \doteq \begin{cases} p(s''|s') & \text{if } \sigma'' = (s', s'') \text{ and } \sigma = (s, s') \\ 0 & \text{otherwise} \end{cases}$$

Or, conversely,

2. To get the setup (6) from that of (43), we keep the state space as is and define $q(y|s, s') = q(y|s')$ for all pairs (s, s'). This formally attaches outputs to transitions while in reality they remain determined solely by the target state s'.

The last remark we wish to make is that the requirement for a unique starting state s_0 is not an essential one. We use it because it simplifies exposition but puts no restriction on our intended use of the HMM concept.

2.10 Additional Reading

A very good heuristic introduction to hidden Markov models can be found in an article by Poritz [7]. It contains an extensive bibliography. For a thorough exposition of the subject, essentially covering the same material as this and the next chapters, see the paper by Rabiner [8].

One problem of concern is the convergence of the estimates derived by the Baum algorithm to the actual HMM parameter values [9]. This is very much an unsolved area, and some thought will reveal why. When HMM estimation is used in speech recognition, convergence is not much of a concern. Other problems (such as overtraining due to sparseness of data, or bad choice of HMM structure or of initial parameter values) dominate. System designers simply carry out three or four iterations and take what comes.

The Baum algorithm attempts to compute maximum likelihood estimates of HMM parameters from provided data. But maximum likelihood

---

11. Note that (6) assumes no null transitions. If we wanted to transform an HMM with null transitions into one with output-producing states, we would need to do it in two steps, the first of which would be the replacement of the null transitions using the formulas of section 2.5.

is surely not the only desirable criterion. How about maximum a posteriori probability, or an estimate that maximizes the mutual information (see chapter 7) between the HMM parameters θ and the observed sequence $y_1, y_2, \ldots, y_k$? Both of these criteria imply the existence of a distribution over the parameter space Θ from which θ is chosen. This is not natural to speech recognition. There are, however, analogous formulations of the speech recognition problem that do involve these alternative criteria. We will return to them in the last section of the next chapter.

References

[1] C.E. Shannon, "A mathematical theory of communication," *Bell System Technical Journal*, vol. 27, pp. 379–423, 623–56, July, October 1948.

[2] W. Feller, *Introduction to Probability Theory and Its Applications*, Vol. I, 3rd ed., J. Wiley, New York, 1968.

[3] J.G. Kemeny and J.L. Snell, *Finite Markov Chains*, Van Nostrand, Princeton, NJ, 1960.

[4] A.J. Viterbi, "Error bounds for convolutional codes and an asymptotically optimum decoding algorithm," *IEEE Transactions on Information Theory*, IT-13, pp. 260–67, 1967.

[5] L.E. Baum and T. Petrie, "Statistical inference for probabilistic functions of finite state Markov chains," *Annals of Mathematical Statistics*, vol. 37, pp. 1559–63, 1966.

[6] L. Baum, "An inequality and associated maximization technique in statistical estimation of probabilistic functions of a Markov process," *Inequalities*, vol. 3, pp. 1–8, 1972.

[7] A.B. Poritz, "Hidden Markov models: a guided tour," *Proceedings of the International Conference on Acoustics, Speech, and Signal Processing*, vol. 1, pp. 7–13, New York Hilton, New York City, April 11–14, 1988.

[8] L.R. Rabiner, "A tutorial on hidden Markov models and selected applications in speech recognition," *Proceedings of the IEEE*, vol. 37, no. 2, pp. 257–86, February 1989.

[9] J.F. Korsh, "Exponential bounds for error and equivocation based on Markov chain observations," *Information and Control*, vol. 22, no. 2, pp. 107–22, 1973.

Chapter 3

The Acoustic Model

3.1 Introduction

In this chapter we will show two methods by which to construct the acoustic model that the recognizer may use to compute the probabilities $P(\mathbf{A}|\mathbf{W})$ for any acoustic data string $\mathbf{A} = a_1, a_2, \ldots, a_m$ and hypothesized word string $\mathbf{W} = w_1, w_2, \ldots, w_n$. As chapter 1 pointed out, these probabilities are needed so the recognizer can search for the desired transcribed word string $\hat{\mathbf{W}}$ defined by

$$\hat{\mathbf{W}} \doteq \arg \max_{\mathbf{W}} P(\mathbf{A}|\mathbf{W})P(\mathbf{W})$$

We will restrict ourselves to finite acoustic data alphabets $\mathscr{A}$ with several hundreds of symbols (200 is standard). We will make the required generalization to real vector alphabets currently used in the most advanced continuous speech recognition systems after we introduce the Expectation-Maximization algorithm in chapter 9.

The acoustic model will be based on the hidden Markov model (HMM) concept introduced in chapter 2. The general approach will be as follows [1] [2] [3]:

The model for a word string $\mathbf{W}$ will be made up of a *concatenation* of models pertaining to the individual words w_i (in a more sophisticated approach these models may be influenced by the *context* w_{i-1} and w_{i+1}, but we will not concern ourselves with this potential complication at this time). The models for the individual words w_i belonging to the basic vocabulary $\mathscr{V}$ will themselves be made up as a concatenation of yet smaller HMMs, the basic building blocks of the acoustic model system. There are many ways to select such building blocks. This chapter will introduce two.

The need for building blocks is obvious: The vocabulary is large (in the ten thousands) and changeable, so it is practically impossible to tailor models separately for individual words. The two types of acoustic models discussed here differ exactly in the nature of their building blocks. The *phonetic* acoustic model is based on an intuitive linguistic concept. On the other hand, the *fenonic* model is completely self-organized from data and at this stage of the discussion is based on whole words. The former model does not take into account intraword context; neither model as presented is capable of dealing with interword context (coarticulation). The consideration of the latter is deferred to chapter 12.

3.2 Phonetic Acoustic Models

The basic way of constructing HMMs for words is as follows:

1. Create a phonetic dictionary [4] [5] [6] for the vocabulary in question, that is, make available the correspondence between each word v and a sequence $\Phi(v) = \varphi_1, \varphi_2, \ldots, \varphi_{l_v}$, of symbols from a predetermined phonetic alphabet ϕ. $\Phi(v)$ is then an encoding of the pronunciation of v and is referred to as the *phonetic base form* of v.

This step involves making a decision about the phonetic alphabet ϕ to be used. Although the international phonetic alphabet (IPA) is prevalent [5], one could use an ordinary dictionary as a guide. For instance, the *American Heritage Dictionary* [7] gives the following pronunciations:

aluminum $= \varepsilon$ l oo m ε n ε m

green $= $ g r $\bar{\text{e}}$ n

Worcester $= $ w $\widetilde{\text{oo}}$ s t ε r

For some words (e.g., *either, the*) that have several fundamentally different valid pronunciations, multiple base forms must be provided. The phonetic alphabet usually distinguishes between stressed and unstressed vowels and includes silence and end-of-word symbols, and its size is of the order of 100.

2. To each symbol of the phonetic alphabet let there correspond an *elementary* HMM with distinguished starting and ending states.

Figure 3.1 shows the transition structure of an appropriate elementary HMM. Note that it generates at least one output symbol and possibly an unlimited number of them.

3. The HMM for a word v is a concatenation of the elementary HMMs specified by the sequence $\Phi(v)$, where the final state of one HMM is con-

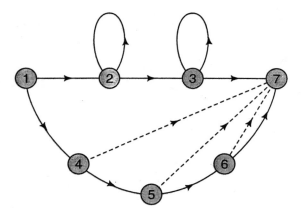

Figure 3.1
Hidden Markov model of acoustic string production by a phone

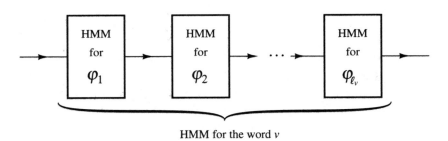

HMM for the word v

Figure 3.2
HMM for the word v as determined by its phonetic baseform

nected to the initial state of the following HMM by a null transition (see figure 3.2).

4. To get a *composite* model for a transcription **W** of some given speech data **A**, word models for the individual words w_i are concatenated by inserting between them elementary HMMs corresponding to silence symbols and/or end-of-word symbols (see figure 3.3).

5. The Baum-Welch algorithm (see section 2.7 of the preceding chapter) estimates the HMM parameters by letting users read a prepared text **W**, observing the acoustic processor's output **A** and using the composite HMM corresponding to **W** as a model of the production mechanism that resulted in the observed **A**.

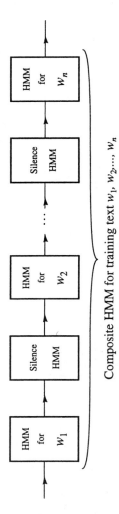

Composite HMM for training text $w_1, w_2, ..., w_n$

Figure 3.3
Composite HMM for the training text $w_1, w_2, \ldots, w_n$

3.3 More on Acoustic Model Training

It may be useful to specify in somewhat more detail what is meant by step 5 in section 3.2. The literal description of HMM training (synonym for estimation of statistical parameters from data) in section 2.7 concerned one HMM that was to account for the production of all the training data **Y** from which the parameters were to be estimated. So, for instance, for each different transition t of the given HMM, one was to estimate the output probability $q(a|t)$. Clearly, this would be possible with any accuracy only if t were used repeatedly as **Y** was being produced.

But this is not necessarily the situation for the composite HMM constructed in step 4 of section 3.2. If all of its transitions were regarded as different, then (in one training iteration) many of them would be used very few times, and most not at all. So this is yet another aspect in which the building block construction is important. During the estimation process, those transitions of the composite HMM that correspond to the same transition in any given building block must be considered the same.[1] Only in this way will most transitions be used many times when the acoustic data **A** is produced.

To carry out training properly we can proceed as follows:

1. Establish an inventory of elementary HMMs. In the rudimentary case described in section 3.2, these correspond to the different phones of the phonetic alphabet ϕ plus the silence and end-of-word HMMs.
2. Give a different index to each different transition of the elementary HMM set. For instance, let $t_{j,k}$ denote the kth transition of the jth building block HMM.
3. For each different $t_{j,k}$ establish different counters $c(t_{j,k})$ and $c(a, t_{j,k})$ to be used as specified in equations (27) and (28) of chapter 2.
4. Identify the composite HMM's transitions by the proper index $t_{j,k}$ of the building block elementary HMM to which they belong.
5. Train the composite HMM as specified in section 2.7, contributing the amounts $P\{t^i = t_{j,k}\}$ and $P\{t^i = t_{j,k}\}\delta(a_{i+1}, a)$ to the $c(t_{j,k})$ and $c(a, t_{j,k})$ counters, respectively:

---

1. That is, if the jth building block appears M times in the construction of the composite HMM, then each transition t of the jth building block will appear in exactly M places in the composite HMM. All M of these will correspond to only one set of accumulators $c(a, t), a \in \mathscr{A}$ for the purpose of estimating the parameters $p(t)$ and $q(a|t)$ of the jth building block.

$$c(t_{j,k}) = \sum_{i=0}^{k-1} P\{t^i = t_{j,k}\}$$

$$c(a, t_{j,k}) = \sum_{i=0}^{k-1} P\{t^i = t_{j,k}\} \delta(a_{i+1}, a)$$

3.4 The Effect of Context

The problem with the phonetic model[2] proposed in section 3.2 is that it does not take into account the influence of context on pronunciation. Thus in American speech an initial t (e.g., in *table*) is usually aspirated, while a final t (e.g., in *nest*) is not. One could try to fix this problem by making the phonetic alphabet allophonic,[3] but that would mean deciding what the allophonic alphabet should be, and hiring a golden-eared (and very conscientious and patient) phonetician to create an allophonic dictionary. We will revisit the allophonic problem in chapter 12 when we derive an allophonic inventory directly from speech data using decision tree methodology.

One way the speech recognition community is attacking the context problem is via *triphones* [8]. This amounts to deciding that the contextual influence of the preceding and the following phones is most important. Or, in allophonic terms, a phone φ is realized by allophones denoted by $x\varphi y$, where x and y range over the phonetic alphabet. The problem with *this* solution is that other, wider contexts may be important and that even in this relatively narrow case, there are potentially K^3 allophones, where K is the size of the phonetic alphabet. For the purposes of HMM parameter estimation this is too many, so triphones must be clustered, and that is again best done via the decision tree methodology discussed in chapter 12.

In this chapter we will solve the phonetic context problem by encoding base forms into a natural, data-driven alphabet that takes into account all intraword context. This solution is thus particularly suitable for isolated word recognition where (short) pauses between words effectively insulate their pronunciation from the surroundings. Compared to the phonetic base form approach of section 3.2, the method will lead to improved speech

2. That is, the problem with the inventory of the building blocks.
3. Allophones are the perceptually different realizations of the same phone [5].

recognition. Furthermore, it will be the basis for our "ultimate" approach to coarticulation (defined as acoustic influence of words on each other) introduced in chapter 12.

We must now make a slight detour and introduce the concept of *Viterbi alignment*.

3.5   Viterbi Alignment

Consider any acoustic data string $\mathbf{A} = a_1, a_2, \ldots, a_m$ and a particular HMM having an inventory of transitions $\mathcal{T}$. *Viterbi decoding* (see section 2.4) refers to finding the most likely sequence of transitions $\mathbf{T}_s = t_{s_1}, t_{s_2}, \ldots, t_{s_k}, (t_{s_i} \in \mathcal{T})$ to have generated $\mathbf{A}$.[4]

The subsequence $\mathbf{T}_o = t_{o_1}, t_{o_2}, \ldots, t_{o_m}$ consisting of the non-null (output producing) transitions of $\mathbf{T}_s$ is the basis of the Viterbi alignment of $\mathbf{A}$ with the HMM. It constitutes a labeling of the symbols a_i of $\mathbf{A}$ by the transitions t_{o_i} that can be thought of as having "caused" the outputs a_i.

If the HMM in question is made up of a concatenation of n elementary HMMs (e.g., the phonetic HMMs of section 3.2 or the word models resulting from their concatenation) then the labeling $\mathbf{T}_o$ effectively segments $\mathbf{A}$ into subsequences of symbols $\mathbf{A}_j^* = a_{l_{j-1}+1}, a_{l_{j-1}+2}, \ldots, a_{l_j}$ that were labeled by transitions belonging to the same elementary HMM. The label $t_{o_i}, i = l_{j-1}$ of the preceding symbol $a_{l_{j-1}}$ (i.e., the corresponding transition) belongs to a different elementary HMM and so does the label $t_{o_h}, h = l_j + 1$ of the succeeding symbol a_{l_j+1} (whereas the transitions $t_{o_{i+1}}, \ldots, t_{o_{h-1}}$, belong to the same elementary HMM).

The term *Viterbi (forced) alignment* refers to this segmentation $\mathbf{A} = \mathbf{A}_1^* \| \mathbf{A}_2^* \| \ldots \| \mathbf{A}_n^*$.

3.6   Singleton Fenonic Base Forms

In spite of the superior performance they facilitate, *fenonic* base forms are not widely used in current speech recognizers. We include their description because they are an excellent illustration of complete self-organization from data. They do not presuppose any phonetic concepts whatever. They are also a useful tool for modeling new words that do not belong to the prepared vocabulary $\mathcal{V}$.

---

4. Because the most likely path may involve null transitions, $k \geq m$.

Having explained Viterbi alignment, we can now address the problem of phonetic context. To establish fenonic base forms [9], an acoustic processor with a finite output alphabet is needed. This does not mean that the method applies only to systems that use such alphabets at run time. The method establishes *word models* from data before any actual recognition takes place. In fact, even in finite alphabet systems it might be useful to have a different acoustic processor at run time from the one used to define these base forms.

Speak a word, say *table*. The output of the acoustic processor is going to be a particular string $\mathbf{A}(table) = a_{i_1}, a_{i_2}, \ldots, a_{i_m}$ with a_{i_j} belonging to the acoustic alphabet $\mathscr{A}$.[5] The string $\mathbf{A}(table)$ is surely characteristic of the pronunciation of the word *table* and can thus be thought of as an *encoding* of that pronunciation.

Speak another word, say *famous*, and consider the consequent acoustic processor output string $\mathbf{A}(famous) = a_{j_1}, a_{j_2}, \ldots, a_{j_l}$. If it should turn out that $a_{i_k} = a_{j_n}$ for some k and n, then that means that the sound of *table* at time k was acoustically similar to the sound of *famous* at time n. It is in this sense that the string $\mathbf{A}(\langle word \rangle)$ constitutes an acoustically faithful encoding of the sound of $\langle word \rangle$.

It thus makes sense to use $\mathbf{A}(\langle word \rangle)$ as a *fenonic* baseform of $\langle word \rangle$, and create the HMM for $\langle word \rangle$ as a concatenation of elementary HMMs corresponding to the individual symbols making up the codeword $\mathbf{A}(\langle word \rangle)$.[6]

Caution: Do not be confused by the fact that the codeword $\mathbf{A}(\langle word \rangle)$ originated as the output string of an acoustic processor! The individual symbols a_{i_h} will be used strictly as abstract identifiers of the particular elementary HMMs whose concatenation will form the composite HMM for $\langle word \rangle$. These identifiers simply serve to identically index acoustically similar segments. The nature of the identifiers' acoustic realization will be determined by training.

---

5. The string $a_{i_1}, a_{i_2}, \ldots, a_{i_m}$ can be obtained from a Viterbi alignment of speech data $\mathbf{A}$ that included the utterance of the word *table*. This alignment can be based on a composite HMM consisting of previously trained *phonetic acoustic word models*. This composite HMM is thought of as having generated $\mathbf{A}$, with the subsequence $a_{i_1}, a_{i_2}, \ldots, a_{i_m}$ aligned with the transitions belonging to the HMM for *table*.

6. The "fe" in the made-up term *fenonic* stands for "front end". The suffix "nonic" is intended to lend the term scientific respectability.

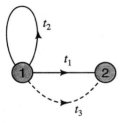

Figure 3.4
Structure of an elementary hidden Markov model of a fenon

Figure 3.4 shows the structure of the elementary HMMs to be used as building blocks. Their statistical parameters will be estimated from data as section 3.3 outlined. Since the length of the fenonic base form is equal to the number of centiseconds (it is assumed that outputs are produced at the rate of 100 per second, but other rates are acceptable) it took to pronounce the word, it is reasonable to expect that after training, the probability $P(t_1)$ of the direct transition in figure 3.4 will turn out to be close to 1 in all the elementary HMMs of the set. The loop and the null transitions in figure 3.4 provide the flexibility that allows the duration of subsequent pronunciations of $\langle word \rangle$ to be different from the one that gave rise to the fenonic baseform.

Different elementary HMMs will have different output probabilities $q(a|t_i)$ associated with the two output producing transitions in figure 3.4. Their values will be derived from training. We will take what comes, but it will not be a coincidence if it should turn out for both non-null transitions t_i belonging to the HMM designated by the *fenonic identifier a* that[7]

$$q(a|t_i) > q(a'|t_i) \qquad \text{for all } a' \neq a$$

3.7   A Needed Generalization

Singleton fenonic base forms have three problems:

1. Base forms are obtained from a single pronunciation of a word that may turn out to have been irregular (coincidental).
2. The base-form creation process requires the user to pronounce each word of a large vocabulary—an impossibly time-consuming task.

---

7. This should hold assuming the same acoustic processor is used at run time as was used to establish fenonic base forms.

3. The base forms do not take into account word context: the influence of the preceding and following words on the pronunciation of the current one.

In fact, it is clear that

1. To avoid atypicality, base forms should be produced from several sample pronunciations of the word.
2. Base forms for a large vocabulary should be produced once and for all and must be the same for each user. Only the statistical parameters $p(t_i)$ and $q(y|t_i)$ can conceivably be speaker dependent.

We are about to see how to construct speaker independent fenonic base forms from multiple recordings of vocabulary words by a variety of speakers. This will obviate the first two problems. The third problem either can be tolerated (this is certainly fine for isolated word recognition) or must be treated in some way similar to that discussed in chapter 12.

Suppose that we have recorded M utterances of some word v, $\mathbf{A}_i(v) = a_{i,1}, a_{i,2}, \ldots, a_{i,m_i}$, $i = 1, 2, \ldots M$. Then the logical thing to seek is a fenonic base form $B^*(v)$ satisfying

$$\mathbf{B}^*(v) = \arg \max_{\mathbf{B}} \prod_{i=1}^{M} P_{\mathbf{B}}(\mathbf{A}_i(v)) \tag{1}$$

In equation (1) the strings $\mathbf{A}_i$ denote ordinary acoustic processor outputs and the subscript to the probability $P_{\mathbf{B}}$ signifies that the HMM producing them is constructed from elementary HMMs specified by the base form $\mathbf{B}$. Of course, this probability is well defined only after the statistical parameters of the HMMs have been estimated, but that is not the main problem.

The unfortunate fact is that the finding of $\mathbf{B}^*$ in (1) involves too large a search space: The duration of an average word exceeds 25 centiseconds, so the number of possible base forms exceeds L^{25} where L is the size of the fenonic alphabet. One possibility is the construction of *synthetic base forms* described in section 3.8. The process will again be based on Viterbi alignment.

3.8   Generation of Synthetic Base Forms

We start the construction by the following training bootstrap:

1. Record all data necessary for speaker independent base form creation for the entire vocabulary $\mathcal{V}$. Each word $v \in \mathcal{V}$ will be recorded by M different speakers.

2. Establish an acoustic processor output alphabet via an appropriate form of vector quantization (see section 1.5) applied to the entire data.

3. Create singleton base forms (by the method of section 3.6) by choosing at random from among the M recordings $\{\mathbf{A}_1(v), \ldots, \mathbf{A}_M(v)\}$ of each word v.

4. Train statistical parameters of the elementary HMMs making up the selected singleton base forms. (This is done on the entire recorded data for all words.)

5. For each word v, choose the "best" new base form $\mathbf{C}^*(v)$ from among the recorded set $\{\mathbf{A}_1(v), \ldots, \mathbf{A}_M(v)\}$:

$$\mathbf{C}^*(v) = \arg \max_{j=1}^{M} \prod_{i=1}^{M} P_{\mathbf{A}_j(v)}(\mathbf{A}_i(v)) \tag{2}$$

6. Using the new base forms $\mathbf{C}^*(v)$, reestimate the statistical parameters of the elementary HMMs (as in step 4).

7. Reselect "best" new base forms (as in step 5) based on the new statistical parameter values obtained in the previous step. Continue in a loop from step 5 until the base-form selection process has converged.

Comparing (2) with (1) we see that whereas we are after the best conceivable base form $\mathbf{B}^*(v)$ giving rise to the data, $\mathbf{C}^*(v)$ is only the best from among the alternative singleton baseforms $\{\mathbf{A}_1(v), \ldots, \mathbf{A}_M(v)\}$. We must therefore push on further.

In the above algorithm we have estimated the statistical parameters of the elementary HMMs that will be used to build up word HMMs according to the specification of the obtained fenonic base forms $\mathbf{C}^*(v)$. We are now ready to construct base forms more nearly satisfying (1) than do the base forms $\mathbf{C}^*(v)$ we selected so far.

The process starts by mutually aligning the symbols of the strings $\mathbf{A}_i = a_{i,1}, a_{i,2}, \ldots, a_{i,m_i}$ [8] as follows:

Using $\mathbf{C}^*$ defined by equation (2) as the base-form specifier for word v, Viterbi-align the symbols of the strings $\mathbf{A}_i$. That is, for each $i = 1, 2, \ldots, M$ find the elementary HMM of the concatenation defined by the base form $\mathbf{C}^*$ that has most likely "produced" the symbol $a_{i,j}$ for each $j = 1, 2, \ldots, m_i$, where m_i is the number of symbols in $\mathbf{A}_i$.

---

8. To avoid complicating our notation unnecessarily, we are dropping the explicit argument v.

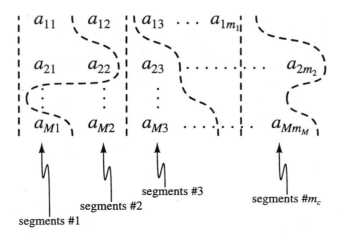

segments #1

segments #2

segments #3

segments #m_c

Figure 3.5
Mutual alignment of acoustic output strings corresponding to M different pronunciations of the same word

If m_C is the length of C^*, the alignment will have segmented each string A_i into m_C segments (some of them possibly of length 0), one segment per symbol of C^*. Figure 3.5 shows an example of this schematically.

Let $r_j(A_i)$ denote the concatenation of the first j segments of the string A_i. In the following, let the inventory of elementary HMMs include the *null HMM* consisting of a single null transition. We now proceed:

1. Find the best concatenation of elementary HMMs, $\mathbf{b}_1^* = b_{1,1}^*, b_{1,2}^*$ producing independently of each other all the first segments of the strings A_i. That is,

$$\mathbf{b}_1^* = \arg \max_{\mathbf{b}_1} \prod_{i=1}^{M} P_{\mathbf{b}_1}(r_1(A_i))$$

where $\mathbf{b}_1 = b_{1,1}, b_{1,2}$ runs over all pairs of elementary HMMs including the null HMM. Note that $\mathbf{b}_1^*$ may in fact consist of the null HMM (the concatenation of two null HMMs is equivalent to a single null HMM), or of one or two "real" (i.e., output producing) elementary HMMs.

2. Find the best HMM pair $b_{2,1}^*, b_{2,2}^*$ to concatenate with $\mathbf{b}_1^*$ so that the resulting HMM specified by $\mathbf{b}_2^* = \mathbf{b}_1^*, b_{2,1}^*, b_{2,2}^*$ is most likely to produce independently the initial segments $r_2(A_i)$. That is, limited by the fixed choice $\mathbf{b}_1^*$ of the initial HMMs, $\mathbf{b}_2^*$ is chosen to maximize the value of the

product

$$\prod_{i=1}^{M} P_{\mathbf{b}_2}(r_2(\mathbf{A}_i))$$

where $\mathbf{b}_2 = \mathbf{b}_1^*, b_{2,1}, b_{2,2}$.

3. Continue in this vein, selecting the HMM pair $b_{j,1}^*, b_{j,2}^*$ so that $\mathbf{b}_j^* = \mathbf{b}_{j-1}^*, b_{j,1}^*, b_{j,2}^*$ will maximize the product

$$\prod_{i=1}^{M} P_{\mathbf{b}_j}(r_j(\mathbf{A}_i))$$

where $\mathbf{b}_j = \mathbf{b}_{j-1}^*, b_{j,1}, b_{j,2}$.

4. The string $\mathbf{B}^* = \mathbf{b}_{m_C}^*$ obtained in this way is the desired synthetic baseform derived from the pronunciations $\mathbf{A}_i = a_{i,1}, a_{i,2}, \ldots, a_{i,m_i}$. Simplify it by eliminating from it all null HMMs.

For best results, the above process should be iterated. That is, once the base forms $\mathbf{B}^*$ are obtained in the last step, the elementary HMMs are retrained on their basis. Then the basic strings $\mathbf{A}_i = a_{i,1}, a_{i,2}, \ldots, a_{i,m_i}$ are aligned with respect to the symbols of $\mathbf{B}^*$ (rather than $\mathbf{C}^*$) and the process starts again from step 1.

3.9 A Further Refinement

Synthetic base form construction can be improved by carrying on in the spirit of the following steps:

1. Find the best concatenation of elementary HMMs, $\mathbf{b}_1^* = b_{1,1}^*, b_{1,2}^*$, $b_{2,1}, b_{2,2}$ producing independently all the first two segments of the strings $\mathbf{A}_i$. That is,

$$\mathbf{b}_1^* = \arg \max_{\mathbf{b}_1} \prod_{i=1}^{M} P_{\mathbf{b}_1}(r_2(\mathbf{A}_i))$$

Note that $\mathbf{b}_1^*$ may in fact consist of the null HMM or of one to four "real" elementary HMMs.

2. Find the best HMM sequence $b_{2,1}^*, b_{2,2}^*, b_{3,1}, b_{3,2}$ to concatenate with $b_{1,1}^*, b_{1,2}^*$ so that the resulting HMM specified by $\mathbf{b}_2^* = b_{1,1}^*, b_{1,2}^*, b_{2,1}^*$, $b_{2,2}^*, b_{3,1}, b_{3,2}$ is most likely to produce independently the initial segments $r_3(\mathbf{A}_i)$. That is, limited by the fixed choice $b_{1,1}^*, b_{1,2}^*$ of the initial HMMs, $\mathbf{b}_2^*$ maximizes the value of the product

$$\prod_{i=1}^{M} P_{\mathbf{b}_2}(r_3(\mathbf{A}_i))$$

3. Continue as in step 2 for subsequent segments.

We need not elaborate the reasons why this improves the resulting synthetic baseform at the price of increased complexity of the search for it.

3.10   Singleton Base Forms for Words Outside the Vocabulary

Finally a remark on the virtues of singleton base forms. They provide an easy means for adding a model of new words to the vocabulary. The user need only say the word and provide its spelling, and the acoustic processor outputs directly specify the required HMM. In fact, this can be done at run time as part of the proofreading of the recognized text. Thus if I said *serendipity* and the system recognized *property* because *serendipity* was not in the vocabulary, I need only make the correction in the recognized text and I gain a singleton fenonic base form for *serendipity* if the recognizer is able to identify (by Viterbi alignment of the speech with the text as currently recognized—see section 3.5) the speech segment that it erroneously recognized as *property*.

3.11   Additional Reading

A very useful review of how to apply HMMs to speech recognition can be found in reference [10]. For a discussion of related practical issues encountered in hidden Markov modeling, see the article by Juang and Rabiner [11].

The inspiration for fenonic word models was no doubt the early work of Bakis [12] who derived HMMs directly from aligned data. His HMM structures are referred to as *Bakis models* and constitute the earliest use of continuous feature vector outputs of the signal processor as inputs to the linguistic decoder.[9]

The mainstream modeling presented in this chapter is based on acoustic processors generating outputs at regular time intervals. This need not be so. Mari Ostendorf and colleagues continue to make fruitful attempts to deal with other natural segmental units [13] [14] [15].

---

9. More on continuous processing will be found in chapter 9.

The Baum algorithm aims at estimating HMM parameters in a maximum likelihood way. That is, it adjusts the parameters of the models of the training script so as to maximize the probability of the models' producing the observed acoustic processor output string. This does not necessarily lead to the best discrimination, that is, speech recognition performance. What one would really wish is to maximize the a posteriori probability $P(\mathbf{W}|\mathbf{A})$ where $\mathbf{W}$ is the spoken word string and $\mathbf{A}$ is the observed acoustic processor output. This turns out to be a rather difficult problem requiring in practice various assumptions and approximations. The first attempts at solution can be found in [16] and [17]. A fundamental step forward was taken by Kanevsky and colleagues who came up with an adjustment to the Baum algorithm that allows training aimed at optimization of the a posteriori probability criterion [18]. A survey of these approaches can be found in [19].

Worth mentioning is also a different attempt to go beyond maximum likelihood in optimizing speech recognizer performance that iteratively adjusts the HMM parameter values so as to make correct and incorrect words more and less probable, respectively [20].

As we mentioned in chapter 1, it is possible to use models other than HMMs as a basis of speech recognition. Among these, artificial neural networks (ANNs) and dynamic time warping (DTW) are preeminent. Bourlard and Wellekens have pointed out interesting connections between HMMs and ANNs [21]. Richard and Lippmann showed (among other properties) that ANNs can be used to estimate a posteriori probabilities [22], and Bourlard and Morgan have taken advantage of this fact in formulating a combined HMM and ANN approach to speech recognition [23]. The system constructed by Robinson, Hochberg, and Renals achieves state-of-the-art results in speaker independent large vocabulary speech recognition [24].

Finally an excursion into the brief history of speech recognition. In the 1970s the dominant paradigm for small vocabulary isolated speech recognition was DTW. Its basic idea was to warp a prototype observation of an utterance into the unknown observed string and reach a decision among competing word candidates according to the warping penalty incurred. A very popular type of this penalty was *Itakura distance* [25]. DTW gave very good results, in fact, for its field of application, better ones than HMMs did (that is, until DTW was essentially abandoned) [26]. The main problems with this approach were (a) incorporation of language models was not natural, (b) the problem of construction of synthetic prototypes remained unsolved,[10] and (c) a unified recognizer statistical formulation incorporating all speech recognizer modules was never found. Two good discussions of DTW applications can be found in references [27] and [28].

References

[1] L.R. Bahl and F. Jelinek, "Decoding for channels with insertions, deletions, and substitutions with applications to speech recognition," *IEEE Transactions on Information Theory*, IT-21, pp. 404–11, 1975.

[2] J.K. Baker, "The Dragon system—an overview," *IEEE Transactions on Acoustics, Speech, and Signal Processing*, ASSP-23, no. 1, pp. 24–29, February 1975.

[3] J.K. Baker, "Stochastic modeling for automatic speech understanding," *Speech Recognition*, D.R. Reddy, ed., Academic Press, New York, 1975.

[4] J.E. Shoup, "Phonological aspects of speech recognition," ch. 6 in *Trends in Speech Recognition*, W.A. Lea, ed., Prentice Hall, Englewood Cliffs, NJ, 1980.

[5] P. Ladefoged, *A Course in Phonetics*, Harcourt Brace Jovanovich, New York, 1975.

[6] P.S. Cohen and R.L. Mercer, "The phonological component of an automatic speech-recognition system," in *Speech Recognition*, D.R. Reddy, ed., Academic Press, New York, 1975.

[7] *The American Heritage Dictionary of the English Language*, Houghton Mifflin, Boston, MA, 1973.

[8] R. Schwartz, C. Barry, Yen-Lu Chow, A. Derr, Ming-Whei Feng, O. Kimball, F. Kubala, J. Makhoul, J. Vandegrift, "The BBN BYBLOS continuous speech recognition system," *Proceedings of the DAPPA Speech and Natural Language Workshop*, pp. 94–99, Morgan Kaufmann Publishers, San Mateo, CA, February 1989.

[9] L.R. Bahl, P. Brown, P. deSouza, R.L. Mercer, and M. Picheny, "Acoustic Markov models used in the Tangora speech recognition system," *Proceedings of the IEEE International Conference on Acoustics, Speech, and Signal Processing*, New York, April 1988.

[10] S.E. Levinson, L.R. Rabiner, and M.M. Sondhi, "An introduction to the application of the theory of probabilistic functions of a Markov process to automatic speech recognition," *Bell System Technical Journal*, vol. 62, no. 4, pp. 1035–74, 1983.

10. The required prototypes had to be constructed from actual samples of word utterances. No smaller building blocks gave adequate performance. Thus the method seemed inapplicable to large vocabulary speech recognition.

[11] B.H. Juang and L.R. Rabiner, " Issues in using hidden Markov models for speech recognition," in *Advances in Signal Processing*, S. Furui and M.M. Sondhi, eds., pp. 509–54, Marcel Dekker, New York, 1992.

[12] R. Bakis, "Continuous-speech word spotting via centisecond acoustic states," IBM Research Report RC 4788, Yorktown Heights, NY, April 2, 1974.

[13] M. Ostendorf and S. Roukos, "A stochastic segment model for phoneme-based continuous speech recognition," *IEEE Transactions on Acoustics, Speech, and Signal Processing*. vol. 37, no. 12, pp. 1857–69, December 1989.

[14] M. Ostendorf, V.V. Digalakis, and O.A. Kimball, "From HMMs to segment models: A unified view of stochastic modeling for speech recognition," *IEEE Transactions on Speech and Audio Processing*, vol. 4, no. 5, pp. 360–78, September 1996.

[15] M. Ostendorf, "From HMMs to segment models: Stochastic modeling for CSR," in *Automatic Speech and Speaker Recognition*, C-H. Lee, F.K. Soong, and K.K. Paliwal, eds., pp. 185–210, Klewer Academic Publishers, Norwell, MA, 1996.

[16] L.R. Bahl, P. Brown, P. deSouza, and R.L. Mercer, "Maximum mutual information estimation of hidden Markov model parameters for speech recognition," *Proceedings of IEEE International Conference on Acoustics, Speech, and Signal Processing*, pp. 49–52, Tokyo, 1986.

[17] S.J. Young, "Competitive training in hidden Markov models," *Proceedings of IEEE International Conference on Acoustics, Speech, and Signal Processing*, pp. 681–84, Albuquerque, NM April 1990.

[18] P.S. Gopalakrishnan, D. Kanevsky, A. Nadas, and D. Nahamoo, "An inequality for rational functions with applications to some statistical estimation problems," *IEEE Transactions on Information Theory*, vol. 37, no. 1, pp. 107–13, January 1991.

[19] Y. Normandin, "Maximum mutual information estimation of hidden Markov models," in *Automatic Speech and Speaker Recognition*, C-H. Lee, F.K. Soong, and K.K. Paliwal, eds., pp. 58–81, Klewer Academic Publishers, Norwell, MA, 1996.

[20] L.R. Bahl, P. Brown, P. deSouza, and R.L. Mercer, "Estimating hidden Markov model parameters so as to maximize speech recognition accuracy," *IEEE Transactions on Speech and Audio Processing*, vol. 1, no. 1, pp. 77–83, January 1993.

[21] H. Bourlard and C.J. Wellekens, "Links between Markov models and multilayer perceptrons," *IEEE Transactions on Pattern Analysis and Machine Intelligence*, vol. 12, no. 6, pp. 1167–78, December 1990.

[22] M. Richard and R. Lippmann, "Neural network classifiers estimate Bayesian à posteriori probabilities," *Neural Computation*, vol. 3, no. 4, pp. 461–83, 1991.

[23] H. Bourlard and N. Morgan, "Hybrid connectionist models for continuous speech recognition," in *Automatic Speech and Speaker Recognition*, C-H. Lee,

F.K. Soong, and K.K. Paliwal, eds., pp. 259–83, Klewer Academic Publishers, Norwell, MA, 1996.

[24] T. Robinson, M. Hochberg, and S. Renals, "The use of recurrent neural networks in continuous speech recognition," in *Automatic Speech and Speaker Recognition*, C-H. Lee, F.K. Soong, and K.K. Paliwal, eds., pp. 233–58, Klewer Academic Publishers, Norwell, MA, 1996.

[25] F. Itakura, "Minimum prediction residual principle applied to speech recognition," *IEEE Transactions on Acoustics, Speech, and Signal Processing*, vol. ASSP-23, no. 1, pp. 67–72, February 1975.

[26] H. Sakoe and S. Chiba, "Dynamic programming algorithm optimization for spoken word recognition," *IEEE Transactions on Acoustics, Speech, and Signal Processing*, vol. ASSP-26, no. 1, pp. 43–49, February 1978.

[27] L.R. Rabiner, A.E. Rosenberg, and S.E. Levinson, "Considerations in dynamic time warping algorithms for discrete word recognition," *IEEE Transactions on Acoustics, Speech, and Signal Processing*, vol. ASSP-26, no. 6, pp. 575–82, December 1978.

[28] C-H. Lee, "Applications of dynamic programming to speech and language processing," *AT&T Technical Journal*, pp. 114–30, Murray Hill, NJ, May/June 1989.

Chapter 4

Basic Language Modeling

4.1 Introduction

It was pointed out in chapter 1 that the task of a recognizer is to find a word string $\hat{\mathbf{W}}$ satisfying

$$\hat{\mathbf{W}} = \arg \max_{\mathbf{W}} P(\mathbf{A}|\mathbf{W})P(\mathbf{W}) \tag{1}$$

where $\mathbf{A}$ denotes the acoustic evidence (data) and

$$\mathbf{W} = w_1, w_2, \ldots, w_n \qquad w_i \in \mathcal{V} \tag{2}$$

denotes a string of n words, each belonging to a fixed and known vocabulary $\mathcal{V}$.

This chapter is our first encounter with *language models* capable of assigning a probability $P(\mathbf{W})$ to every conceivable word string $\mathbf{W}$. We repeat from chapter 1 that the designation *word* refers to a *word form* defined by its spelling. Two differently spelled inflections or derivations of the same stem are considered different words. Homographs having different parts of speech or meanings constitute the same word.

Using the Bayes' rule of probability theory, $P(\mathbf{W})$ can be formally decomposed as

$$P(\mathbf{W}) = \prod_{i=1}^{n} P(w_i|w_1, \ldots, w_{i-1}) \tag{3}$$

where $P(w_i|w_1, \ldots, w_{i-1})$ is the probability that w_i will be spoken given that words $w_1, \ldots, w_{i-1}$ were said previously. The past $w_1, \ldots, w_{i-1}$ is frequently referred to as *history* and is denoted succinctly by $\mathbf{h}_i$.

Formula (3) simply states that the probability of uttering a word string $\mathbf{W}$ is given by the probability of uttering the first word, times the probability of uttering the second word given that the first word was uttered, etc., times the probability of uttering the last word of the string given that all of the previous ones were uttered. Hence in (3), the choice of w_i is modeled to depend on the entire past history of the discourse.

The task of a language model is to make available to the recognizer adequate estimates of the probabilities $P(w_i|w_1, \ldots, w_{i-1})$. In this chapter we introduce the basic and surprisingly powerful *trigram language model*, which remains the state of the art. We show how to compute its parameters directly from training data: text appropriate to the recognition task at hand. Here we will introduce the technique of optimum linear smoothing by the method of held-out estimation that is useful wherever data is sparse.[1]

Another way of combating data sparseness is by equivalence classification of words. The simplest method that comes to mind works according to the grammatical function the words play, for instance by assigning to each word of text a label from a finite set $\mathscr{T}$ of parts of speech (tags). $\mathscr{T}$ would include such conventional elements as NOUN, VERB, ADJECTIVE, ADVERB, DETERMINER, PREPOSITION, CONJUNCTION, etc. It may be quite large, containing also many refinements of the conventional seven parts of speech just listed. It may even have *semantic labels* such as COMPANY NAME, CITY NAME, DATE, PRICE, etc.

In English, words spelled the same way may fulfill different grammatical functions depending on the context in which they are used. For instance, the word *light* used as a NOUN may denote a device that illuminates, as a VERB the action of illuminating, and as an ADJECTIVE the qualities of absence of either darkness or weight. In any case, most English NOUNS can be used as VERBS and vice versa. Assignment of parts of speech, called *tagging*, can be done by Viterbi decoding based on appropriate HMMs, as section 4.8 will show. In section 4.9 we will demonstrate a possible incorporation of tag equivalence classes into the language model. In section 4.10 we will consider the problem of vocabulary selection for a particular recognition task.

We will return four more times to language modeling, in chapters 10, 13, 14, and 15. In those chapters, we will discuss considerably more sophisticated methods than those introduced here.

---

1. Section 4.7 briefly discusses the quite prevalent but more complex method of backing-off [1], but a derivation of the method is delayed to chapter 15.

4.2 Equivalence Classification of History

The sequential decomposition (3) is appropriate for speech recognition because it leads to the development of natural intermediate decision criteria allowing a minimal delay in the recognizer's response to the progressing dictation. In reality, of course, the probabilities $P(w_i|w_1, \ldots, w_{i-1})$ would be impossible to estimate for even moderate values of i, since most histories $w_1, \ldots, w_{i-1}$ would be unique or would have occurred only a very few times. Indeed, for a vocabulary of size $|\mathcal{V}|$ there are $|\mathcal{V}|^{i-1}$ different histories, and so to specify $P(w_i|w_1, \ldots, w_{i-1})$ completely, $|\mathcal{V}|^i$ values would have to be estimated. For practical values of $|\mathcal{V}|$, this is an astronomically large number (for instance, for $|\mathcal{V}| = 5000$ and $i = 3$, $|\mathcal{V}|^i$ is equal to 125 billion!), and thus the probabilities $P(w_i|w_1, \ldots, w_{i-1})$ could neither be stored nor retrieved when the recognizer needed them.

It follows that the various possible conditioning histories $w_1, \ldots, w_{i-1}$ must be distinguished as belonging to some manageable number M of different equivalence classes [2] [3]. Indeed, from a common sense point of view, it is obvious that thinking of many of the histories as equivalent will not appreciably weaken the model's ability to predict the next word w_i. Besides, for practical vocabulary sizes (in the tens of thousands) and $i > 3$, the vast majority of the $|\mathcal{V}|^i$ word strings $w_1, \ldots, w_i$ will never occur in English.

Let Φ be a (many to one) mapping of histories into some (perhaps large) number M of equivalence classes. If $\Phi(w_1, \ldots, w_{i-1})$ denotes the equivalence class of the string $w_1, \ldots, w_{i-1}$, then the probability $P(\mathbf{W})$ may be approximated by

$$P(\mathbf{W}) = \prod_{i=1}^{n} P(w_i|\Phi(w_1, \ldots, w_{i-1})) \tag{4}$$

One way the classifier might function is on the basis of a finite state "grammar." At time $i - 1$, the grammar is in state Φ_{i-1}, and the next word forces a change to state Φ_i. Then (4) can be re-written as

$$P(\mathbf{W}) = \prod_{i=1}^{n} P(w_i|\Phi_{i-1}) \tag{5}$$

For this relatively simple situation, how would the probabilities $P(w_i|\Phi_{i-1})$ be estimated?

One could acquire some large corpus of text of the kind that the recognizer will produce (e.g., if the application is the creation of medical reports, then the training text should consist of medical reports composed by any available method). One would then run the text's word sequences through the finite state grammar and accumulate counts $C(w, \Phi)$ of the number of times the word w was fed to the grammar immediately after the grammar was in state Φ. If $C(\Phi)$ denotes the number of times the grammar reached state Φ,

$$C(\Phi) = \sum_w C(w, \Phi) \tag{6}$$

then the first-order estimate of the desired probability would be

$$P(w_i | \Phi_i = \Phi) = \frac{C(w_i, \Phi)}{C(\Phi)} \tag{7}$$

Intuitively, a "grammatical" classification of a history $w_1, \ldots, w_{i-1}$ might be something like

w_{i-1} was *the*; it is a part of a verb phrase with head *blows*; the previous phrase was a noun phrase with head *exhaust*.

Whatever the selected classification scheme, it will of necessity be a compromise between two requirements:

1. The classification must be sufficiently refined to provide adequate information about the history **h** so it can serve as a basis for prediction.
2. When applied to histories of the given training corpus, it must yield its M possible classes frequently enough so that the probabilities $P(w | \Phi)$ can be reliably estimated (not necessarily by the crude relative frequency approach (7)).

The purpose of a language model for speech recognition is not an exact analysis for meaning extraction, but an apportionment of probability among alternative futures. This remark points to the possibility of creative use of relatively simple grammatical principles that are on the whole accurate, even if open to counterexamples.

4.3   The Trigram Language Model

The language model most frequently used by present-day speech recognizers is based on a very simple equivalence classification: *Histories are equivalent if they end in the same two words.* For such a *trigram model*,

$$P(\mathbf{W}) = \prod_{i=1}^{n} P(w_i|w_{i-2}, w_{i-1}) \tag{8}$$

Estimating the basic trigram probabilities by formulas (6) and (7) gives

$$P(w_3|w_1, w_2) = f(w_3|w_1, w_2) \doteq \frac{C(w_1, w_2, w_3)}{C(w_1, w_2)} \tag{9}$$

where $f(\ |\ )$ denotes the relative frequency function. Unfortunately, (9) is a very inadequate formula, since many possible English word trigrams w_1, w_2, w_3 are never actually encountered even in very large corpora of training text. A language model based on (9) would assert the impossibility of such trigrams and would assign $P(\mathbf{W}) = 0$ to strings $\mathbf{W}$ containing them. A recognizer operating under the statistical decision criterion (1) would thus be forced to commit a large number of errors. In fact, in the 1970s IBM researchers conducted an experiment in which they divided a corpus of patent descriptions based on a 1,000-word vocabulary into test and training subsets (of size 300,000 and 1,500,000 words, respectively). They found that 23 per cent of the trigrams appearing in the test subset never occurred in the training subset. Thus a language model based on formula (9) would have caused the recognizer to transcribe erroneously at least 23 per cent of the spoken words.

It is therefore necessary to *smooth* the trigram frequencies [3]. This can be done most simply by interpolating trigram, bigram, and unigram relative frequencies,

$$P(w_3|w_1, w_2) = \lambda_3 f(w_3|w_1, w_2) + \lambda_2 f(w_3|w_2) + \lambda_1 f(w_3) \tag{10}$$

where the nonnegative weights satisfy $\lambda_1 + \lambda_2 + \lambda_3 = 1$.[2] We will next discuss how to choose the weights λ_i optimally.

---

2. There are other ways of smoothing. In fact, nonparametric probability estimates extracted from data, though dependent on event counts, need not be based on relative frequencies at all. Many formulas have been suggested that do not assign 0 probability to unseen events. The most famous among them is the Good-Turing formula [4]. Chapter 15 discusses this and other approaches. Until that chapter, to simplify our discussion, we will use relative frequencies. But it should be kept in mind that this is only a pedagogical decision. Whatever formulas derived from counts are used, linear smoothing (discussed in the next section) can be applied to them.

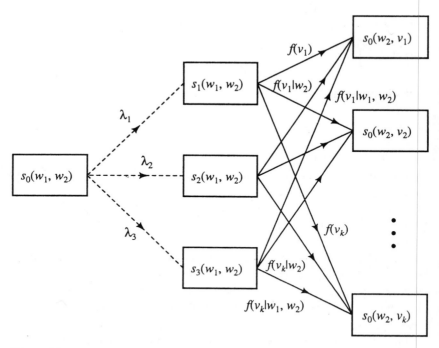

Figure 4.1
Linear smoothing section of an HMM language model

4.4 Optimal Linear Smoothing

A language model operating according to (10) can be regarded as an HMM [5]. Figure 4.1 shows one of its subparts. Outgoing from its leftmost state $s_0(w_1, w_2)$ are three null transitions t_1, t_2, and t_3 into states $s_1(w_1, w_2)$, $s_2(w_1, w_2)$, and $s_3(w_1, w_2)$. The transitions are taken with probabilities λ_1, λ_2, and λ_3, respectively. Out of each of the three states lead $|\mathcal{V}|$ transitions. Each produces a different output $v \in \mathcal{V}$ and accordingly leads to a state $s_0(w_2, v)$. The v-transitions out of states $s_1(w_1, w_2)$, $s_2(w_1, w_2)$, and $s_3(w_1, w_2)$ are taken with probabilities $f(v), f(v|w_2)$, and $f(v|w_1, w_2)$, respectively.

Note that the probabilities of the output-producing transitions are known, whereas those of the null transitions are to be determined. The overall HMM is, of course, gigantic: it has $4 \times |\mathcal{V}|^2$ states. In agreement with (10), however, for a given $i \in \{1, 2, 3\}$ all probabilities λ_i of transitions leading from $s_0(w_1, w_2)$ to $s_i(w_2, v)$ are to have the same value regardless of the actual combination w_1, w_2, v. We say that their values are *tied*.

Since the language model (10) is an HMM, the Baum algorithm (section 2.7) can estimate the optimal λ_i values.[3] As a welcome consequence of tying the probabilities of null transitions, the estimation process will need only three counters $c(t_i)$. In fact, regardless to which subpart of the HMM a transition marked t_i belongs, the same counter $c(t_i)$ will receive as training data contributions the probabilities (denoted in chapter 2 by $P\{t^j = t_i\}$) that the transition t_i was taken. We will not need to establish counters for the output-producing transitions because their probabilities are fixed.

Additional practical shortcuts can be taken to make estimating the weights λ_i quite easy. We will discuss these in section 4.6. Right now we ask the important question: What type of training data should be used for determining the weights?

It cannot be the data on which the relative frequencies $f(\mid)$ were computed because in that case the resulting estimates would be $\lambda_3 = 1$, $\lambda_1 = \lambda_2 = 0$. In fact (as is proven in section 4.6), $f(w_3|w_1, w_2)$ is the *maximum likelihood estimate* of $P(w_3|w_1, w_2)$ for the training data on which $f(w_3|w_1, w_2)$ is based. Looking at the problem intuitively, the HMM includes the branch t_1 so that the word w_3 can be "generated" (we use quotation marks because the HMM does not in reality generate any text) even when the bigram w_2, w_3 has not been found in the training data, and similarly t_1 and t_2 make "generation" of w_3 possible when the trigram w_1, w_2, w_3 is absent.

So we conclude that the total training data must be divided into two portions. The first, much larger, called *kept* data, is first used to estimate the relative frequencies $f(\mid)$. With these fixed, the second smaller data portion (because there are very many fewer parameters to estimate), called *held-out* data, is used to estimate the weights λ_i. Of course, once this is done, the language model can be further improved by combining both portions of the data and reestimating the $f(\mid)$'s.

This technique of linear smoothing is sometimes called *deleted interpolation*. To introduce it in its simplest setting, we treated the weights λ_i as constants. However, since it is clear that $f(w_3|w_1, w_2)$ approximates $P(w_3|w_1, w_2)$ better if based on a larger count $C(w_1, w_2)$, then λ_is should

---

3. Actually, it is more practical to base the estimation directly on calculus, as is done in section 4.6.

themselves depend on the conditioning counts $C(w_1, w_2)$ and $C(w_2)$. This does not violate the HMM formulation, it merely changes the tying of its null transitions. One simple way to accomplish this is to change the basic HMM component structure from figure 4.1 to that of figure 4.2. There we have added a state $s_4(w_1, w_2)$ to the HMM subpart and as a consequence have changed the set of null transitions. Now transitions t_3 and t_4 go from state $s_0(w_1, w_2)$ to states $s_3(w_1, w_2)$ and $s_4(w_1, w_2)$, whereas transitions t_1 and t_2 leave state $s_4(w_1, w_2)$ and go to states $s_1(w_1, w_2)$ and $s_2(w_1, w_2)$. Denoting the new transition probabilities by λ_i' we have the relation

$$\lambda_1 = \lambda_4' \times \lambda_1'$$

$$\lambda_2 = \lambda_4' \times \lambda_2' \tag{11}$$

$$\lambda_3 = \lambda_3'$$

Of course, $\lambda_2' = 1 - \lambda_1'$ and $\lambda_4' = 1 - \lambda_3'$.

The point of the structure of figure 4.2 is that it smoothes $P(w_3|w_1, w_2)$ in two steps.[4] First we get

$$P^*(w_3|w_2) = \lambda_1' f(w_3) + \lambda_2' f(w_3|w_2) \tag{12}$$

and then

$$P(w_3|w_1, w_2) = \lambda_4' P^*(w_3|w_2) + \lambda_3' f(w_3|w_1, w_2) \tag{13}$$

By the argument of the paragraph preceding (11), the parameters λ_i' in (12) should depend on the counts $C(w_2)$. Then (12) yields the best obtainable estimate (by the deleted interpolation method) of $P^*(w_3|w_2)$, and in (13) we interpolate this P^* with the relative frequency $f(w_3|w_1, w_2)$ and thus get the "best" estimate of $P(w_3|w_1, w_2)$. Of course, in (13) the parameters λ_i' depend on the counts $C(w_1, w_2)$. As a result of all this we see from (11) that the overall λ_i parameters (used in the structure of figure 4.1) will indeed depend on both counts $C(w_1, w_2)$ and $C(w_2)$, as they should.

The Baum algorithm applied to the HMM based on figure 4.2 can then estimate the values of λ_i' by appropriately tying the null transitions.[5] That is, all transitions t_3 (and t_4) that emanate from such states $s_0(w_1, w_2)$ as correspond to the same count $C(w_1, w_2)$ are tied together, and all tran-

4. This method of smoothing is top-down because it generates higher- in terms of lower-order estimates. Bottom-up smoothing is also possible.

5. Section 4.6 shows a more efficient method of estimation.

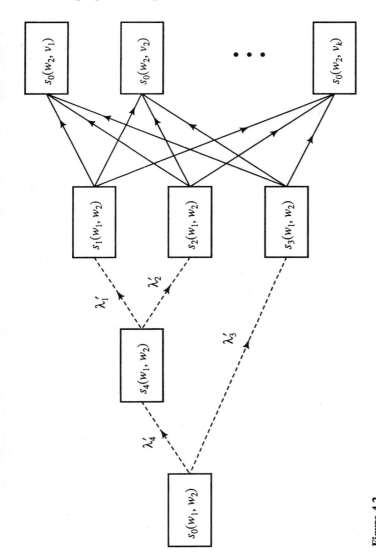

Figure 4.2
Two step, top-down estimation of weights of *n*-gram relative frequencies

sitions t_1 (and t_2) that emanate from such states $s_4(w_1, w_2)$ as correspond to the same count $C(w_2)$ are tied together.

We will return to the smoothing problem in section 4.6, where we discuss smoothing's practical aspects.

4.5   An Example of a Trigram Language Model

Examination of figure 4.3 may provide an intuitive feeling for the quality of the language model (12) and (13). The language model is used to predict, in their order, the words of the sentence *We need to resolve all the important issues within the next two days*. The figure displays all words predicted to be at least as likely as the actual word, given the language model's perfect knowledge of the past. (At the beginning of the sentence it knows nothing except that the sentence begins.) Thus, knowing the two preceding words *all the*, the language model estimates that the most likely next word is *necessary*, and that the words *data*, ..., *shop* are all more likely than the actual word *important* which it estimates as the 1641st likeliest, given the particular past *all the*. We can observe that this language model is quite good at predicting most function words (e.g., *we*, *to*, *the*, *next*, etc.) but uncertain about some content words (e.g., *resolve* and *important*) just as human beings might be. (No claim is made here that they wouldn't do better, just that they too might have trouble.)

A careful look at figure 4.3 reveals one of deleted interpolation's shortcomings. Note that the word *the* appears in the second column as one that follows more frequently the sequence $w_1 = \text{SB}$, $w_2 = We$ (SB stands for *sentence beginning*) than does *need*. Of course, the sequence *We the* is not completely impossible in English (the U.S. Constitution starts with *We the people* ...), but it is surely less probable than *We need*. The reason for the unnaturally high probability of *We the* is the presence of the component $\lambda'_1 \times f(w_3 = the)$ in the right-hand side of (12) and the fact that *the* is the most probable English word.

Of course, one example proves nothing. We would like to have an objective measure of language model quality. The results of Information theory, with which we will deal in chapter 6, suggest such a measure to us.

4.6   Practical Aspects of Deleted Interpolation

Section 4.3 treated the problem of language model smoothing from a rigorous HMM point of view. It is immediately obvious, however, that

1	The	are	**to**	know	the	issues	necessary	
2	This	will		have	this	problems	data	
3	One	the		understand	these	**the**	information	
4	Two	would		do	problems		above	
5	A	also		get	any		other	
6	Three	do		the	a		time	
7	Please	**need**		use	problem		people	
8	In			provide	them		operators	
9	**We**			insert	**all**		tools	
·				·			·	
·				·			·	
·				·			·	
93				request			factors	
94				respond			facts	
95				supply			I	
96				write			jobs	
97				me			MVS	
98				**resolve**			old	
·							·	
·							·	
·							·	
1636							mailroom	
1637							marketplace	
1638							provision	
1639							reception	
1640							shop	
1641							**important**	

1	role	and	**the**	**next**	be	meeting	of	
2	thing	from			**two**	months	·	
3	that	in				years		
4	to	to				meetings		
5	contact	are				to		
6	parts	with				week		
7	point	were				**days**		
8	for	requiring						
9	**issues**	still						
·		·						
·		·						
·		·						
61		being						
62		during						
63		I						
64		involved						
65		would						
66		**within**						

Figure 4.3
Quality of the language model's prediction of the sentence *We need to resolve all the important issues within the next two days*

the actual estimation can be broken down to make the computation much simpler than the formal HMM approach would demand. We start with the requirement that the smoothing weights be functions of occurrence counts $C(w_1, w_2)$ and $C(w_2)$. As a result, (12) can be changed to

$$P^*(w_3|w_2) = \gamma(C(w_2)) \times f(w_3) + (1 - \gamma(C(w_2))) \times f(w_3|w_2) \tag{14}$$

and (13) to

$$
\begin{aligned}
P(w_3|w_1, w_2) = {} & \theta(C(w_1, w_2)) \times P^*(w_3|w_2) \\
& + (1 - \theta(C(w_1, w_2))) \times f(w_3|w_1, w_2)
\end{aligned}
\tag{15}
$$

The γ coefficients of (14) can be estimated first and independently for each different value of the count $C(w_2)$. Actually, the γ values should depend only on *ranges* into which $C(w_2)$ falls, if for no other reason than because very few words w_2 will have high values of $C(w_2)$. So for the purpose of this section, let $\mathscr{R}(w_2)$ denote the range to which the count $C(w_2)$ belongs. The ranges would be determined experimentally, making sure that enough data falls into them. Of course, for low count values the ranges might contain a single count only, whereas the last range would contain all counts exceeding some threshold.

The coefficients γ for (14) would be estimated as follows for a particular range $\mathscr{R}$ (estimations for different ranges are independent of each other):

1. Divide the total training data into kept and held-out data sets (see section 4.3).
2. Compute the relative frequencies $f(w_3|w_2)$ and $f(w_3)$ from the kept data.
3. Compute $N(w_2, w_3)$, the number of times the bigram w_2, w_3 takes place in the held-out data set.[6]
4. Find γ maximizing the value[7]

$$\sum_{N(v) \in \mathscr{R}} \sum_{w_3} N(v, w_3) \log[\gamma \times f(w_3) + (1 - \gamma) \times f(w_3|v)] \tag{16}$$

Now $\gamma = \gamma(\mathscr{R})$ can, of course, be found by the Baum algorithm's re-estimation process. But actually (16) is so simple that it can be solved faster with the help of calculus [6].

---

6. Note that we designate by $C(\ )$ the counts in the kept training text, and by $N(\ )$ the counts in the held-out set.

7. It is easy to see that (16) corresponds to the maximum likelihood criterion.

Taking the derivative with respect to γ, performing some simple arithmetic, and equating the result to 0, we get

$$\sum_{N(v) \in \mathcal{R}} \sum_{w_3} N(v, w_3) \left[\gamma + \frac{f(w_3|v)}{f(w_3) - f(w_3|v)} \right]^{-1} = 0 \qquad (17)$$

The equation (17) has a single solution because the second derivative of (16) is equal to

$$- \sum_{N(v) \in \mathcal{R}} \sum_{w_3} N(v, w_3) \left[\gamma + \frac{f(w_3|v)}{f(w_3) - f(w_3|v)} \right]^{-2}$$

which is a negative quantity for all γ. The solution to (17) can be found by any appropriate interval search [7].

We can now prove easily the assertion we made in section 4.3 that if we had tried to find γ from the same data we used to compute the relative frequencies $f(w_3|w_2)$, we would have gotten the trivial answer $\gamma = 0$. Indeed, in that case

$$C(v, w_3) = f(v) f(w_3|v) C$$

where C is the size of the development data set. Replacing $N(v, w_3)$ in (17) by $C(v, w_3)$ we get

$$C \sum_{C(v) \in \mathcal{R}} f(v) \sum_{w_3} f(w_3|v) \left[\frac{f(w_3) - f(w_3|v)}{\gamma (f(w_3) - f(w_3|v)) + f(w_3|v)} \right]$$

and setting $\gamma = 0$ and simplifying by cancellation, we end up with

$$C \sum_{C(v) \in \mathcal{R}} f(v) \sum_{w_3} [f(w_3) - f(w_3|v)]$$

But the inner sum is equal to 0 for all v, and so in this case $\gamma = 0$ is a solution to (17); since (17) has a single solution, then $\gamma = 0$ is its only solution.

We have now shown a practical way to find the coefficients $\gamma(\mathcal{R})$. Once these are found, they determine the probability $P^*(w_3|w_2)$ of (14). With the latter known, finding the values $\theta(\mathcal{R}(w1, w2))$ for (15) is the same problem as that of finding γ's for (14).

4.7 Backing-Off

Linear smoothing is not the only way to make up for the scarcity of text training data. The method of backing-off [1] is quite prevalent in state-of-the-art speech recognizers. We will postpone its derivation to chapter 15, in which we introduce Good-Turing estimation [4]. The basic idea is apparent from the following formula:

$$\hat{P}(w_3|w_1, w_2) = \begin{cases} f(w_3|w_1, w_2) & \text{if } C(w_1, w_2, w_3) \geq K \\ \alpha Q_T(w_3|w_1, w_2) & \text{if } 1 \leq C(w_1, w_2, w_3) < K \\ \beta(w_1, w_2)\hat{P}(w_3|w_2) & \text{otherwise} \end{cases} \quad (18)$$

where α and β are appropriately chosen so that the probability $\hat{P}(w_3|w_1, w_2)$ is properly normalized. Furthermore, $Q_T(w_3|w_1, w_2)$ is a Good-Turing type function and $\hat{P}(w_3|w_2)$ is a bigram probability estimate having the same form as $\hat{P}(w_3|w_1, w_2)$:

$$\hat{P}(w_3|w_2) = \begin{cases} f(w_3|w_2) & \text{if } C(w_2, w_3) \geq L \\ \alpha Q_T(w_3|w_2) & \text{if } 1 \leq C(w_2, w_3) < L \\ \beta(w_2)f(w_3) & \text{otherwise} \end{cases} \quad (19)$$

The form (18) then constitutes a recursion. The thresholds K and L are chosen intuitively.

The argument for backing-off is that if there is enough evidence for it (if the trigram count $C(w_1, w_2, w_3)$ is sufficiently large), then relative frequency is a very good estimate of the probability. If not, then one should back-off and rely on bigrams; if there is not enough evidence even for those, then one has to make do with unigrams.

4.8   HMM Tagging

As pointed out in section 4.1, words spelled the same way may fulfill different grammatical functions depending on the context in which they are used. In general, therefore, to each word there will correspond a set of tags (parts-of-speech functions) that it can assume. As long as the tag set $\mathcal{T}$ is more or less conventional, we can find the possible word-tag correspondences in commercial dictionaries.

The basic idea of HMM tagging derives from the following model of text production[8] [8]: Each tag $g \in \mathcal{T}$ can be realized as the word $v \in \mathcal{V}$ with probability $k(v|g)$. (Naturally, we will set $k(v|g) = 0$ if v cannot have the tag g.) The model produces text statistically by first generating a string

---

8. It cannot be stressed enough that the model's only purpose is to facilitate tagging. It has no other plausibility as a method of text generation.

of tags $G = g_1g_2 \ldots g_n$ and then converting them into the words $w_i, i = 1, 2, \ldots, n$ with probability $k(w_i|g_i)$.

If $P(g_i|g_1, \ldots, g_{i-1}) = h(g_i|g_{i-k}, \ldots, g_{i-1})$, then we have an HMM whose states are the tag k-grams $g_{i-k} \ldots g_{i-1}$. We can then tag by reversing the production process and answering the question:

Given that the observed word sequence $w_1w_2 \ldots w_n$ was produced by the HMM model, which was the most likely tag sequence $g_1g_2 \ldots g_n$ underlying $w_1w_2 \ldots w_n$?

The answer to the question is obtained by Viterbi decoding (see section 2.4).

HMM tagging is usually based on a trigram model of tag production, $P(g_i|g_1, \ldots, g_{i-1}) = h(g_i|g_{i-2}, g_{i-1})$. We must therefore determine the basic model probabilities $k(w_i|g_i)$ and $h(g_i|g_{i-2}, g_{i-1})$.

The most straightforward way is with human help. A training text is hand-tagged and the smoothed[9] relative frequencies observed in the tagged training text become the desired probabilities k and h. The tedious work of tagging can be accelerated by feedback:

1. A moderate amount of text is tagged by humans and the initial parameter values for k and h are determined.
2. More text is Viterbi tagged, based on the obtained k and h values.
3. Human annotators correct the Viterbi tagging.
4. New k and h values are calculated on the basis of *all* the text tagged.
5. If step 3 did not yield the desired accuracy, repeat from step 2; else stop.

Of course, one would like to estimate k and h without human intervention, or at least with very little of it. This has been tried with moderate success [9] [10] based on the Baum-Welch algorithm initialized by

$$h(g_i|g_{i-2}, g_{i-1}) = \frac{1}{|\mathcal{T}|} \tag{20}$$

$$k(v|g) = \frac{k^*(g|v)f(v)}{\sum_{v'} k^*(g|v')f(v')} \tag{21}$$

where

---

9. Linear smoothing analogous to that of section 4.4 may be used; that is,
$$h(g_i|g_{i-2}, g_{i-1}) = \lambda_3 f(g_i|g_{i-2}, g_{i-1}) + \lambda_2 f(g_i|g_{i-1}) + \lambda_1 f(g_i)$$

$$k^*(g|v) = \begin{cases} \dfrac{1}{L(v)} & \text{if } g \text{ is a possible tag of } v \\ 0 & \text{if } g \text{ is not a possible tag of } v \end{cases} \tag{22}$$

In (22) $L(v)$ denotes the number of possible tags of the particular word v.

Unfortunately, no matter how much data it is based on, the Baum-Welch determination of model parameters leads to worse tagging results than if k and h has been obtained from human text annotation (assuming an adequate amount of annotated text). The problem is not due to bad initialization. In fact, if we initialize with the best values of k and h obtainable by hand tagging and run the Baum-Welch algorithm on more data, the new parameter values will give worse results than the initial ones [10].

The problem lies with the maximum likelihood optimization criterion that underlies Baum-Welch. It is not directly related to our actual aim of maximizing tagging accuracy, and so it does not lead to it.

4.9 Use of Tag Equivalence Classification in a Language Model

The simplest way to introduce a tag component into a language model would be to tag training text and then collect relative frequency statistics $f(v|g_1, g_2)$ and $f(v|g)$ for all combinations $v \in \mathscr{V}$ and $g, g_1, g_2 \in \mathscr{T}$. One would then create a smoothed probability estimate

$$Q(v|g_1, g_2) = \gamma f(v|g_1, g_2) + (1 - \gamma)f(v|g_2)$$

and interpolate it with the trigram language model of the form (10)

$$\hat{P}(w_i|w_{i-2}, w_{i-1}) = \mu P(w_i|w_{i-2}, w_{i-1}) + (1 - \mu)Q(w_i|g(w_{i-2}), g(w_{i-1})) \tag{23}$$

The problem with (23) is that its implied tagging function $g(w_i)$ does not exist, or put another way, depends on all of the words $w_1, \ldots, w_i, \ldots, w_n$ of the utterance. (This is how the Viterbi algorithm operates.) The straightforward, rigorous solution is to replace the Q probability in (23) by

$$Q^*(w_i|w_1, \ldots, w_{i-1}) \doteq \sum_{g_1, g_2} Q(w_i|g_1, g_2) \tag{24}$$
$$P(g(w_{i-2}) = g_1, g(w_{i-1}) = g_2|w_1, \ldots, w_{i-1})$$

where $P(g(w_{i-2}) = g_1, g(w_{i-1}) = g_2|w_1, \ldots, w_{i-1})$ is the corresponding forward probability of the HMM with states (g_1, g_2) based on the probabilities $k(w_i|g_i)$ and $h(g_i|g_{i-2}, g_{i-1})$. That is, in the terminology of section 2.3,

$$P(g(w_{i-2}) = g_1, g(w_{i-1}) = g_2|w_1, \ldots, w_{i-1}) = \frac{\alpha_i(g_1, g_2)}{\sum_{g'_1, g'_2} \alpha_i(g'_1, g'_2)} \qquad (25)$$

Obviously, the need for the calculation of the probabilities $Q^*(w_i|w_1, \ldots, w_{i-1})$ at recognition time is unfortunate, and so it is natural to seek an equivalence classification that assigns to each word a unique category. One such classification can be found in section 10.12 [11]. Others are discussed in section 4.11.

4.10 Vocabulary Selection and Personalization from Text Databases

Every speech recognizer is based upon a vocabulary $\mathcal{V}$ of finite size $|\mathcal{V}|$ that includes only the words that can appear in the transcribed text. As a consequence, any attempt to recognize a word not belonging to $\mathcal{V}$ will result in an error. Such an error will increase the probability of an error in recognizing the next word. On the other hand, since the difficulty of constructing a language model and its complexity grow with the vocabulary size, it is desirable that the vocabulary be as small as possible. Therefore, to accommodate both needs, the vocabulary must be chosen carefully so that the expected coverage of the dictated text (the percentage of words in the running text that belong to $\mathcal{V}$) is maximized while the total number of words it contains is kept manageable.

It is necessary to measure the coverage appropriately. There is a marked difference between self-coverage and coverage. In the early 1980s, the research group at IBM examined a corpus of office correspondence and found that its most frequent 5,000 words covered 98 per cent of the corpus, but their recognizer based on that vocabulary was committing a surprising number of errors. Further examination showed that the chosen vocabulary covered new text (produced in the same offices) only 92.5 per cent of the time. The difference between self-coverage and coverage is further highlighted by the following: (a) There are only 29,000 distinct words in all of the works of Shakespeare, and (b) Perusal of *The American Heritage Word Frequency Book* [12] shows that quite familiar words such as *admonition* or *deluded* are not among the 86,000 most frequent words. (And, of course, technical words belonging to any particular field, proper names, address words, etc., will not be found in *The American Heritage Word Frequency Book* at all.)

A vocabulary is particular to a speaker and his task in at least four different aspects. First, choice of words reflects the speaker's habits of expression that are related, for example, to his level of education. His usage is also conditioned by the general domain of discourse (e.g., data processing, musicology, medical reports, etc.) that calls for a variety of technical terms, clichés, and such. Furthermore, the speaker has available those names, addresses, and other expressions peculiar to his current interests. Finally, he requires words specific to the topic of the particular document he is composing.

A large, fixed vocabulary can at most provide for the speaker's habitual usage and for the discourse domain, but a completely satisfactory coverage can be attained only through a user's active, though limited, participation. In particular, although he can be asked to indicate whether a proper noun is a personal or company name, location, etc., and even to annotate technical terms, he cannot be expected to provide any grammatical feature classification of ordinary English words. Information about the latter should either be contained in some large stored source lexicon, or, better still, be extracted from the actual syntactic/semantic function that these words play in the previously transcribed text.

Using a rather sophisticated vocabulary selection method, R.L. Mercer of IBM constructed vocabularies of varying sizes. Table 4.1 shows their coverage of the PDB corpus (correspondence of a rather prolific member of the IBM research staff) when names and acronyms occurring in it had been disregarded.

A crude method of personalization could be devised by maintaining during dictation a dynamically varying vocabulary consisting at any given moment of the last L different words used. The resulting *dynamic coverage* is then defined as the asymptotic probability that the word spoken next already belongs to the vocabulary. Table 4.2 gives the result for the PDB database. The second column is the good news: 99 per cent coverage

Table 4.1

Vocabulary size	Static text coverage (%)
5,000	92.5
10,000	95.9
15,000	97.0
20,000	97.6

Table 4.2

Vocabulary size	Dynamic text coverage (%)	Text size needed to reach coverage
5,000	95.5	56,000
10,000	98.2	240,000
15,000	99.0	640,000
20,000	99.5	1,300,000

is achieved by a 15,000-word vocabulary. The third column is the bad: It takes 640,000 words of text to assemble a set of 15,000 different words. This then may be a practical way of personalization only if the vocabulary is extracted from a single pool of texts produced by many users belonging to the same organization.

For direct personalization it would be desirable if a request for a low-frequency word brought automatically into the vocabulary other words that collocate with the requested word. Perhaps a very large "dormant" vocabulary could be classified into overlapping word sets, and the use of a certain number of such words belonging to the same set would trigger the activation of the rest of the set, with consequent deactivation of unused sets. At the start of the process, the recognizer would be provided with an active vocabulary consisting of a permanent core subset of the most frequent English words (derived from general databases), and of an initial, replaceable, subset of words used in one of a number of specified business fields.

4.11   Additional Reading

Language model estimation is necessarily a sparse data problem, and so smoothing techniques become crucial to it. Section 4.4 introduced linear smoothing. Ney and Essen [13] and Essen and Steinbiss [14] discuss the problem from several fundamental points of view.

A trigram language model is built up from trigram, bigram, and unigram relative frequencies (see equations (10), (18), and (19)). Is it useful to attempt n-gram language models with $n > 3$? The answer is in general no, especially because the slight improvement in perplexity would be dearly paid for by the model's considerably increased complexity. However,

Guyon and Pereira [15] have developed a construction technique[10] that keeps complexity within prescribed limits and yet allows for use of n-grams with an (in principle) unlimited value of n. A slight improvement over this approach and some alternatives are presented in reference [16].

A generalization of the concept of trigrams has been achieved by Lafferty and colleagues [17]. Basing their technique on link grammars [18], they are able to take into account nonlocal influences in the text.

As mentioned in sections 4.8 and 4.9, it is desirable (for instance to limit the language model size, but also for smoothing purposes) to base at least some components of a language model on an equivalence classification of words. Based on (multiple) part-of-speech classification, this has been done for Italian [19]. Reference [11] is based on unique class assignment of words. The basic technique used is that described in section 10.12. The classification is derived automatically from data and depends on the use to which words are put and the predictive relation between them. Using simulated annealing [20], Jardino improves on this method [21] [22] by allowing multiple classification of words,[11] thus complicating the language model (as seen in section 4.9). Equivalence classification can of course also be based directly on semantics [23].

Finally, the concept of a word defined as a sequence of letters separated by space should not be sacred. Shouldn't collocations like *New York* or *nuclear magnetic resonance* be treated as single items? Smadja's technique [24] can be used to determine which expressions are worthy of single word treatment.

References

[1] S. Katz, "Estimation of probabilities from sparse data for the language model component of a speech recognizer," *IEEE Transactions on Acoustics, Speech and Signal Processing*, vol. 35, no. 3, pp. 400–01, March 1987.

[2] F. Jelinek, "Self-organized language modeling for speech recognition," in *Readings in Speech Recognition*, A. Waibel and K.F. Lee, eds., pp. 450–506, Morgan-Kaufmann, San Mateo, CA, 1990.

10. The technique depends on the concept of *divergence* (section 7.3) and is based on decision trees, the subject of chapter 10.

11. Obviously, words spelled the same way but having different grammatical functions (see the above example of *light*) or radically different meanings (e.g., *bass* the fish and the musical instrument) should be allowed entrance into different categories.

[3] F. Jelinek, R.L. Mercer, and S. Roukos, "Principles of lexical language modeling for speech recognition," in *Advances in Speech Signal Processing*, S. Furui and M.M. Sondhi, ed., Marcel Decker, New York, 1991.

[4] I.J. Good, "The population frequencies of species and the estimation of population parameters," *Biometrika*, vol. 40, parts 3 and 4, pp. 237–64, December 1953.

[5] F. Jelinek and R.L. Mercer, "Interpolated estimation of Markov source parameters from sparse data," *Proceedings of the Workshop on Pattern Recognition in Practice*, pp. 381–97, North Holland, Amsterdam, 1980.

[6] L.R. Bahl, P.F. Brown, P.V. de Souza, R.L. Mercer, and D. Nahamoo, "A fast algorithm for deleted interpolation," *Proceedings of Eurospeech 91*, pp. 1209–12, Genova, Italy, September 1991.

[7] W.H. Press, S.A. Teukolsky, W.T. Vetterling, and B.P. Flannery, *Numerical recipes in C*, 2nd ed., pp. 350–54, Cambridge University Press, Cambridge, 1992.

[8] L.R. Bahl and R.L. Mercer, "Part of speech assignment by a statistical algorithm," IEEE International Symposium on Information Theory, Ronneby, Sweden, June 1976.

[9] A.M. Derouault and F. Jelinek, "Modèle probabiliste d'un langage en reconnaissance de la parole," *Annales de telecommunications*, tome 39, no. 3–4, pp. 143–51, mars–avril 1984.

[10] B. Merialdo, "Tagging English text with a probabilistic model," *Computational Linguistics*, vol. 20, no. 2, pp. 155–72, June 1984.

[11] P.F. Brown, V.J. Della Pietra, P.V. deSouza, J.C. Lai, and R.L. Mercer, "Class-based *n*-gram models of natural language," *Computational Linguistics*, vol. 18, no. 4, pp. 467–80, December 1992.

[12] J.B. Carroll, P. Davies, B. Richman, *The American Heritage Word Frequency Book*, Houghton Mifflin, Boston, MA, 1971.

[13] H. Ney and U. Essen, "On smoothing techniques for bigram-based natural language modeling," *Proceedings of the IEEE International Conference on Acoustics, Speech and Signal Processing*, pp. 825–28, Toronto, May 1991.

[14] U. Essen and V. Steinbiss, "Cooccurrence smoothing for stochastic language modeling," *Proceedings of the IEEE International Conference on Acoustics, Speech and Signal Processing*, vol. I, pp. 161–64, San Francisco, CA, March 1992.

[15] I. Guyon and F. Pereira, "Design of a linguistic postprocessor using variable memory length Markov models," *Proceedings of the Third ICDAR*, pp. 454–57, Montreal 1995.

[16] J. Hu, W. Turin, and M.K. Brown, "Language modeling with stochastic automata," *Proceedings of the International Conference on Spoken Language Processing*, vol. 1, pp. 406–09, Philadelphia, PA, October 1996.

[17] S. Della Pietra, V. Della Pietra, J. Gillett, J. Lafferty, H. Printz, and L. Ureš, "Inference and estimation of a long-range trigram model," in *Grammatical Inference and Applications*, Lecture notes in Artificial Intelligence 862, R.C. Carrasco and J. Oncina, eds., pp. 78–82, Springer Verlag, Berlin 1994.

[18] D. Grinberg, J. Lafferty, and D. Sleator, *A Robust Parsing Algorithm for Link Grammars*, Technical Report CMU-CS-95-125, School of Computer Science, Carnegie-Mellon University, Pittsburgh, PA, August 1995.

[19] G. Maltese, F. Mancini, "An automatic technique to include grammatical and morphological information in a trigram-based statistical language model," *Proceedings of the IEEE International Conference on Acoustics, Speech and Signal Processing*, vol. I, pp. 157–60, San Francisco, CA, March 1992.

[20] S. Kirkpatrick, C.D. Gellat, and M.P. Vecchi, "Optimization by simulated annealing," *Journal of Statistical Physics*, vol. 34, no. 5–6, pp. 975–86, March 1984.

[21] M. Jardino and G. Adda, "Automatic determination of a stochastic bigram class language model," in *Grammatical Inference and Applications*, R.C. Carrasco and J. Oncina, eds., Lecture Notes in Artificial Intelligence 862, pp. 57–65, Springer Verlag, Berlin 1994.

[22] M. Jardino, "Multilingual stochastic *n*-gram class language models," *Proceedings of the IEEE International Conference on Acoustics, Speech and Signal Processing*, vol. I, pp. 161–63,. Atlanta, GA, May 1996.

[23] G.A. Miller, R. Beckwith, C. Fellbaum, D. Gross, and K.J. Miller, "Introduction to Wordnet: an on-line lexical data base," *Journal of Lexicography*, vol. 3, no. 4, pp. 235–44, 1990.

[24] F. Smadja, "Retrieving collocations from text: xtract," *Computational Linguistics*, vol. 19, no. 1, pp. 143–77, March 1993.

Chapter 5

The Viterbi Search

5.1 Introduction

We have now discussed, at least on a basic level, acoustic processing, acoustic modeling, and language modeling. The remaining component of a speech recognizer is the hypothesis search (see section 1.2). Several methods are possible; the ones we discuss in this chapter are based on the Viterbi algorithm [1] introduced in sections 2.4 and 2.5.[1] Tree search methods will be presented in the next chapter.

5.2 Finding the Most Likely Word Sequence

We will now describe a method of search for the most likely word sequence $\mathbf{W}$ to have caused the observed acoustic data $\mathbf{A}$. This method, like all the known others for large vocabulary sizes $|\mathscr{V}|$, cannot actually be guaranteed to find the most likely $\mathbf{W}$. It gives very good results, however, from the practical point of view (i.e., it is very rare that the maximizing sequence

$$\hat{\mathbf{W}} = \arg \max_{\mathbf{W}} P(\mathbf{A}|\mathbf{W})P(\mathbf{W}) \tag{1}$$

is the one actually spoken but is not found by our search method).[2]

---

1. This will demonstrate again the usefulness of the HMM formulation. We will take further advantage of the fact that embedding HMMs into an HMM leads to a new HMM. In a sense, the subject of this chapter is already known to the reader even before we delve into it.

2. As we know from chapter 2, the Viterbi algorithm (which is the basis of our method) finds the most likely path through an HMM. But (1) demands that we find for each candidate word string $\mathbf{W}$ the probability of the set of paths that

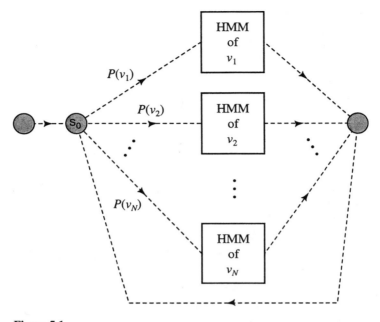

Figure 5.1
Composite model of speech generation when word generation is assumed to be memoryless

First of all, consider the simplest language model, one in which all histories $w_1, w_2, \ldots, w_{i-1}$ are equivalent (see section 4.1). Then

$$P(\mathbf{W}) = \prod_{i=1}^{n} P(w_i) \tag{2}$$

It is immediately obvious that in this case finding $\hat{\mathbf{W}}$ amounts to searching for the most likely path through the graph of figure 5.1. This statement must be interpreted literally—we don't care what happens inside the HMM boxes.

correspond to that $\mathbf{W}$, and then identify the word string whose set of paths has the highest probability. Fortunately, in practice it is very rare that the word string corresponding to the most probable set of paths is the one actually spoken but the one corresponding to the most probable path is not. How to estimate the error rate due to the Viterbi approximation and other shortcuts is discussed in section 6.8.

The slightly more complicated bigram language model assigns probabilities by the formula

$$P(\mathbf{W}) = P(w_1) \prod_{i=2}^{n} P(w_i|w_{i-1}) \tag{3}$$

Then $\hat{\mathbf{W}}$ is given by the most likely path through the graph of figure 5.2 (those of its null transitions not labeled by probability expressions have probability 1). Note that while figure 5.2 is somewhat more complicated than figure 5.1, the number of acoustic models (HMMs) in both is the same. The total number of states is proportional to the vocabulary size $|\mathscr{V}|$. Thus, except in cases where estimates of $P(v_i|v_j)$ cannot be obtained, one would always use the bigram model and thus conduct the search through the graph of figure 5.2 rather than that through figure 5.1.[3]

For a trigram language model,

$$P(\mathbf{W}) = P(w_1)P(w_2|w_1) \prod_{i=3}^{n} P(w_i|w_{i-2}, w_{i-1}) \tag{4}$$

the graph is considerably more complex—the number of states is proportional to $|\mathscr{V}|^2$. For $|\mathscr{V}| = 2$, figure 5.3 illustrates the transition structure. There we have marked the multiple outgoing transitions by the *generation probabilities* to which they correspond. In general, similar graphs can be derived for any equivalence classifier Φ, as long as the number of equivalence classes is finite.

How do we find the most likely paths through these graphs? No practical algorithms exist for finding the exact solution. However, if we replace the boxes in figures 5.1 through 5.3 by the corresponding HMM models, then all the figures simply represent (huge) HMMs in their own right. The Viterbi algorithm will therefore find the most likely path through such HMMs. This path will lead through a sequence of word models whose identity can then specify the recognized word string.[4]

5.3 The Beam Search

The only problem with the Viterbi approach is that for practical vocabularies (e.g., $|\mathscr{V}| = 20,000$) even the bigram HMM represented in figure 5.2

---

3. This is because the language model (3) is more accurate than model (2).

4. We have already explained in footnote 2 why this does not necessarily correspond to the string $\hat{\mathbf{W}}$ defined by (1).

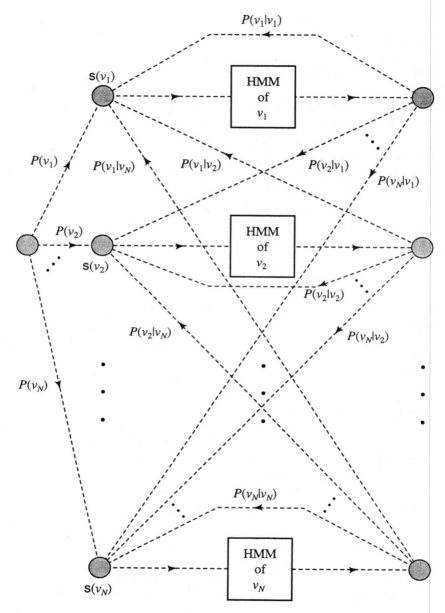

Figure 5.2
Composite model of speech generation when the generated words are assumed to depend only on the identity of the preceding word (bigram language model)

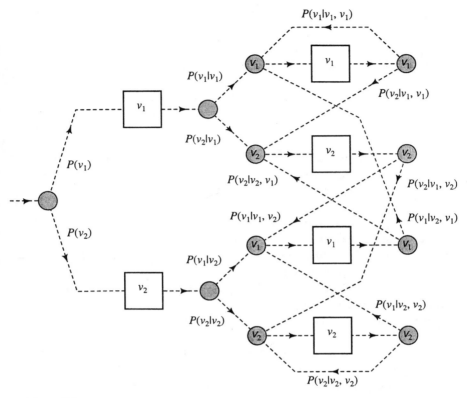

Figure 5.3
A section of the transition structure of speech generation based on a trigram language model when the vocabulary consists of two words only

has too many states. One short cut is the so-called *beam search* [2]. The idea is simple. Imagine the trellis (see section 2.3) corresponding to figure 5.2. Before carrying out the Viterbi purge (step 4 of the algorithm described in section 2.4) at stage i, we determine the maximal probability P_{i-1}^m of the states at stage $i - 1$

$$P_{i-1}^m \doteq \max_s \gamma_{i-1}(s) \tag{5}$$

In (5) $\gamma_{i-1}(s)$ is defined in chapter 2

$$\gamma_i(s) \doteq \max_{s_1,\ldots,s_l} P(s_1, s_2, \ldots, s_{l-1}, s_l = s, y_1, y_2, \ldots, y_i | s_0) \tag{6}$$

and is computed by recursion (18) of chapter 2, repeated here for ready reference:

$$\gamma_i(s) = \max\left\{\max_{s'} r(y_i, s|s')\gamma_{i-1}(s'), \max_{s''<s} q(s|s'')\gamma_i(s'')\right\} \tag{7}$$

The value P_{i-1}^m serves as a basis for a dynamic threshold

$$\tau_{i-1} = \frac{P_{i-1}^m}{K} \tag{8}$$

where K is a suitably chosen constant.[5] We then eliminate from the trellis all states s' on level $i - 1$ such that

$$\gamma_{i-1}(s') < \tau_{i-1} \tag{9}$$

That is, in the recursion (7) for the next level i we set $\gamma_{i-1}(s') = 0$ for all states s' that satisfy (9).

This purge of improbable paths reduces drastically the number of states entering the comparison implied by the max function in the recursion (7) without significantly affecting the values $\gamma_i(s)$ (if K is appropriately chosen), and makes the resulting beam search algorithm a practical one, at least for the case of bigram language models.

5.4   Successive Language Model Refinement Search

Is there any practical way to implement the search for $\hat{\mathbf{W}}$ for a trigram language model? One ingenious method has been suggested by researchers at SRI [3]. We will describe the main idea.

The HMMs for each word in the vocabulary have a final state (see the far right column of states in figure 5.2). Final states of words, therefore, occur at various stages of the trellis. As the beam search proceeds, some of these states will be killed by the thresholding (9). Consider now those word final states in the trellis that remain alive. A trace-back (see Viterbi algorithm step 5 in section 2.4) from any of these final states, say from word v at (time) stage j, will lead to the initial state of the same word model, say at stage i. The pair (i, j) then identifies a time interval during which it is reasonable to admit the hypothesis that the word v might have been spoken.

---

5. The value of the constant K must be determined experimentally for each different application. A good start is $K \approx 100$.

To each word $v \in \mathcal{V}$ there will then correspond a set of time intervals $\{(i_1(v), j_1(v)), (i_2(v), j_2(v)), \ldots, (i_m(v), j_m(v))\}$ during which it might have been uttered. Some of these intervals will overlap since the beam search may admit many ending times for a word actually uttered at a certain instant.[6] Other intervals will represent the presence of the word at different positions in the discourse. Yet others will reflect "false alarms."

We can hypothesize that word v' could conceivably follow word v if and only if there exist intervals $(i_k(v), j_k(v))$ and $(i_l(v'), j_l(v'))$ such that $i_l(v') \in (i_k(v), j_k(v))$ and $j_k(v) \in (i_l(v'), j_l(v'))$. We can, therefore, construct a directed graph whose intermediate states correspond to words $v \in \mathcal{V}$ and are accordingly marked by v. This graph will have two properties:

1. Any path from the initial to the final stage of the graph will pass through states corresponding to a word sequence $w_1, w_2, \ldots, w_n$ such that the word interval sets permit w_i to follow w_{i-1}.

2. Arcs leading to a state corresponding to a word v' emanate only from states marked by the same word v. Put another way, if arcs from states marked v and v'' lead to a state marked v' then *necessarily* $v = v''$.[7]

To the arcs of this graph we can then attach trigram probabilities $P(w_i | w_{i-2}, w_{i-1})$, and we can expand the states of the graph by replacing them with the HMMs of the corresponding word. We can then construct a new trellis for the resulting overall HMM. Compared to a trellis constructed directly for the trigram language model (cf. figure 5.3), the new trellis will have many fewer states.

Recapitulating, to find $\hat{\mathbf{W}}$ we conduct two successive beam searches. The first, on the trellis of the bigram HMM, results in time intervals indicating the possible presence of particular words. The knowledge of these intervals gives rise to a new HMM, associated with trigram probabilities. A second beam search is then conducted over the trellis of this second HMM and results in the final choice of $\hat{\mathbf{W}}$.

---

6. These intervals will certainly overlap with intervals corresponding to other words $v' \in \mathcal{V}$.

7. This property assures that the transitions out of a state corresponding to v' can be associated with a unique trigram probability: Given any state s associated with v', all previous states are necessarily associated with the same word v, and so the identity of s determines the next word distribution $P(w_i | w_{i-1} = v, w_{i-2} = v')$. Figure 5.3 satisfies this property if its boxes are interpreted as states. For instance, it is possible to reach the first (second) state marked v_1 in the next-to-last column only by being previously in either of the two states marked v_1 (v_2).

Obviously, this two-pass strategy can be extended further if better and better, but more complex, language models are available. After each pass a new transition graph is prepared, allowing for the application of the next-order language model.

5.5   Search Versus Language Model State Spaces

So far have assumed that the language model used determines the complexity of the composite HMM. So the schematic structure of figure 5.2 was necessitated by a bigram language model, and that of figure 5.3 by a trigram model. If such rigidity could not be relaxed then a Viterbi-based search would be practically impossible for language models whose predictive power stems from the consideration of events in a more distant past. The only way to accommodate these models would be via the previous section's multi-pass strategy.

Fortunately this is not the case. We are perfectly free to adopt a history equivalence classification for the composite HMM that differs from that for the language model.[8] So, for instance, we are free to use the structure of figure 5.2 with a trigram language model: Observe that once the search reaches a particular stage of the corresponding trellis, the Viterbi decision assures that only a single path will lead into each state of that stage. Thus the language model probability of the next word can depend on the entire past, starting at the initial stage. So although it is true that the composite HMM of figure 5.2 will require a decision among all of the live paths ending with the same word $w_i = v$ (which is suboptimal for a trigram language model), these decisions can be based on probabilities $P(w_i = v | w_{i-2} = v', w_i = v'')$ for all live combinations (v', v''), and indeed on probabilities $P(w_i = v | \mathbf{h}_i = \mathbf{v})$ for all past live paths $\mathbf{v}$.

5.6   *N*-Best Search

In many applications it is desirable to find the N most likely word sequences given the observed acoustic data **A**. For instance, we may wish to reprocess the data using more refined models whose complexity is such that the basic recognizer could not have used them directly. These models

---

8. We do not claim that this would be optimal, only that it might be expedient.

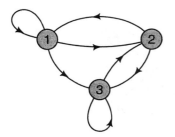

Figure 5.4
Simple HMM used to illustrate two-best search

may be acoustic, providing a sharper estimate of $P(\mathbf{A}|\mathbf{W})$, or linguistic, resulting in a better (or simply an additional) estimate of $P(\mathbf{W})$. As an instance of the latter case, the various alternate hypotheses $\mathbf{W}_1, \mathbf{W}_2, \ldots, \mathbf{W}_N$ could be *parsed*[9] or processed to yield their meaning. This postprocessing would then resolve the remaining N-fold ambiguity.

The required N-best search would differ from the Viterbi search in one aspect only: Instead of retaining only one path leading into each trellis state, each trellis state is split into N states, one for each of the N most likely paths leading into the unsplit state from the (split) states of the trellis's previous stage. This is illustrated for $N = 2$ in figure 5.5 based on the simple HMM of figure 5.4:

• Figure 5.5a is a diagram of a stage of the usual trellis corresponding to the HMM of figure 5.4.
• In figure 5.5b the "entry" states of the trellis stage are split in two, representing the two best paths into each entry state. To each of the split states would normally be attached the probability of the single path leading into it. The transitions out of the split states into the next stage are identical within each pair.
• Into the "exit" states of the trellis of figure 5.5b there now lead multiple paths, of which only the two most likely (for each state) should be

9. By parsing we mean supplying a phrase structure analysis of the sentence. This is usually done by means of a parse tree whose nodes are annotated by phrase markers (such as NOUN PHRASE, VERB PHASE, PREPOSITIONAL PHRASE, etc.) identifying the type of the word sequence found at the leaves of the subtree emanating from the annotated node.

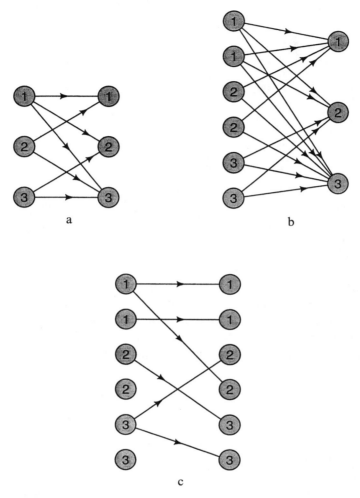

Figure 5.5
Illustration of two-best search based on the HMM of figure 5.4
Figure 5.5a
Trellis stages corresponding to the HMM of figure 5.4
Figure 5.5b
Entry states of the trellis stage of figure 5.5a split so that the identically marked entry states correspond to the best and second-best paths into the corresponding state of the stage of figure 5.5a
Figure 5.5c
Exit states of the stage of figure 5.5b split so that the identically marked exit states correspond to the best and second-best paths into the corresponding state of the stage of figure 5.5b

retained. This is illustrated in figure 5.5c where the exit states of figure 5.5b are appropriately split.

• The split exit states of figure 5.5c become the entry states of the next stage, which will again have the appearance of Figure 5.5b.

If the above modification of the Viterbi algorithm is carried out, the best N sequences are found by determining the N most probable (split) states of the last trellis stage and by retracing the trellis paths leading to them.

5.7 A Maximum Probability Lattice

In section 5.4, describing a search based on successive refinement of the language model, we have encountered a structure commonly referred to as a *lattice* (we didn't use that term). The lattice indicated which words were likely to have been spoken at which time intervals. We can create a lattice in yet another way based on a maximum a posteriori probability approach.

Let $\mathscr{S}(v)$ denote the set of states of the composite HMM (e.g., that of figure 5.2) which are internal to the HMM for word v. We can then define the probability

$$P^*\{s^i \in \mathscr{S}(v)\} \doteq \text{the probability that } \mathbf{A} = a_1, a_2, \ldots, a_m \text{ was produced}$$
$$\text{and at the } i^{th} \text{ trellis stage the HMM was in one of}$$
$$\text{the states of } \mathscr{S}(v).$$

It follows directly from the discussion of section 2.7 that

$$P^*\{s^i \in \mathscr{S}(v)\} = \sum_{s \in \mathscr{S}(v)} \alpha_i(s)\beta_i(s)$$

We can therefore determine, for each (time) stage i of the trellis, the set of the l most probable words $\mathscr{V}_i = \{v_{i,1}, v_{i,2}, \ldots, v_{i,l}\}$ in the process of being uttered at that time i. The number l can either be a fixed constant and

$$P^*\{s^i \in \mathscr{S}(v_{i,j})\} \geq \max_{v \notin \mathscr{V}_i} P^*\{s^i \in \mathscr{S}(v)\}, j = 1, 2, \ldots, l$$

or it can be determined by use of a suitably chosen threshold τ:

$$v_{i,1} = \arg \max_v P^*\{s^i \in \mathscr{S}(v)\}$$

$$\mathscr{V}_i = \{v : P^*\{s^i \in \mathscr{S}(v)\} \geq P^*\{s^i \in \mathscr{S}(v_{i,1})\} \times \tau\}$$

The sets $\{v_{i,1}, v_{i,2}, \ldots, v_{i,l_i}\}$, $i = 1, 2, \ldots, m$ then constitute a time-aligned map of words whose utterance was a likely cause of the observed sequence **A**. This is the basis of the maximum probability lattice we are after.

Obviously, to produce such a lattice we will have to cope with the computational problems presented by the enormity of the composite HMM that corresponds to the recognition task. We will thus need to use thresholding methods analogous to those that defined the beam search of section 5.3.

5.8   Additional Reading

Another multiple-pass search (section 5.4) is presented in reference [4]. Ney and Aubert consider the general problem of dynamic programming searches [5].[10] Finally Kogan considers the problem of early decision in Viterbi decoding [6].[11]

References

[1] A.J. Viterbi, "Error bounds for convolutional codes and an asymmetrically optimum decoding algorithm," *IEEE Transactions on Information Theory*, vol. IT-13, pp. 260–67, 1967.

[2] B.T. Lowerre, *The Harpy Speech Recognition System*, Ph.D. Dissertation, Department of Computer Science, Carnegie-Mellon University, Pittsburgh, PA, 1976.

[3] H. Murveit, J. Butzberger, V. Digalakis, and M. Weintraub, "Large-vocabulary dictation using SRI's Decipher Speech Recognition System: Progressive Search Techniques," *Proceedings of the Spoken Language Systems Technology Workshop*, Massachusetts Institute of Technology, Cambridge, MA, January 1993.

[4] R. Schwartz, L. Nguyen, and J. Makhoul, "Multiple-pass search strategies," in *Automatic Speech and Speaker Recognition*, C-H. Lee, F.K. Soong, and K.K. Paliwal, eds., pp. 429–56, Klewer Academic Publishers, Norwell, MA, 1996.

10. Viterbi decoding can be shown to be a variant of dynamic programming search.

11. The way we described it in chapter 2, the trace-back that determines the decoding decision is carried out only when the entire utterance was processed by the algorithm. The question is whether the speech recognizer could make its initial decisions earlier, and if so, what conditions must be met.

[5] H. Ney and X. Aubert, "Dynamic programming search: from digit strings to large vocabulary word graphs," in *Automatic Speech and Speaker Recognition*, C-H, Lee, F.K. Soong, and K.K. Paliwal, eds., pp. 385–413, Klewer Academic Publishers, Norwell, MA, 1996.

[6] J.A. Kogan, "Exact Viterbi recognition of hidden Markovian sequences through the most informative stopping times," in *Image Models (and Their Speech Model Cousins)*, S.E. Levinson and L. Shepp, eds., pp. 115–30, Springer-Verlag, New York, 1996.

Chapter 6

Hypothesis Search on a Tree and the Fast Match

6.1 Introduction

In speech recognition, having observed the acoustic processor string **A** we wish to find the word string $\hat{\mathbf{W}}$ that is most likely to have "caused" **A**:

$$\hat{\mathbf{W}} = \arg \max_{\mathbf{W}} P(\mathbf{A}|\mathbf{W})P(\mathbf{W}) \tag{1}$$

Since $\mathbf{W} = w_1, w_2, \ldots, w_n$, we can think of the problem as that of searching for a path in a tree whose branches are labeled with the various words of the vocabulary $\mathcal{V} = \{v_1, v_2, \ldots, v_{|\mathcal{V}|}\}$.[1] Such a tree has $|\mathcal{V}|$ branches leaving each node, one for each word. Figure 6.1 illustrates this point of view for $|\mathcal{V}| = 3, n = 4$.

In this chapter we will consider search methods seeking a path through a tree of possible hypotheses. Because in speech recognition, a hypothesis search concerns trees with many branches (of the order of tens of thousands) leaving a node, we will be forced to limit the computational effort by employing a method, called *fast match*, that will reject from consideration the vast majority of these branches without subjecting them to detailed analysis. The Viterbi search of chapter 5 achieved a similar limitation by the combined action of thresholding (which led to the beam search) and multipass processing.

The idea of tree searches is that while looking for the best complete path, it should be necessary to evaluate only relatively few of the potential complete paths, the rest being abandoned (*pruned*) early (the search is

---

1. $|\mathcal{V}|$ denotes the size of the vocabulary.

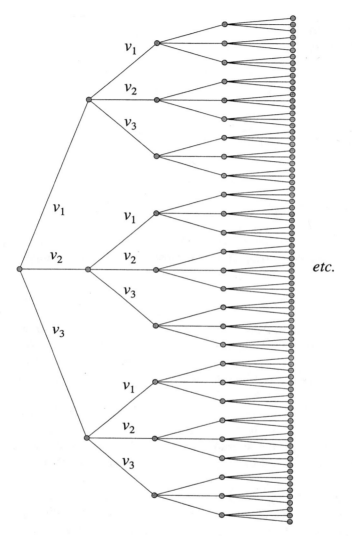

etc.

Figure 6.1
Hypothesis tree for a vocabulary of three words

from the root of the tree outward) because such paths possess some observable (measurable) characteristic making them unlikely to be best. In this chapter we will introduce the desired multistack algorithm [4] [5] in a series of steps designed to make its development plausible and to place it in the context of developments in artificial intelligence and information theory.

6.2 Tree Search Versus Trellis (Viterbi) Search

A Viterbi search of a trellis, described in chapter 5, is an attempt to find the most likely succession of transitions through a composite HMM made up of word HMMs and (null) transitions between them. In the composite HMM, the end states of the word HMMs correspond to the equivalence classes $\Phi(w_1, \ldots, w_{k-1})$ of language model histories. The number of states in a trellis stage is proportional to the number of these end states.[2] The latter then must be limited to keep the Viterbi search's computational and storage requirements feasible. The tree search we are about to describe imposes no such limitation on the number of composite HMM end states, and that is one of its advantages (as long as this search does not prune out the correct path).

We further remind the reader of footnote 2 of chapter 5, where we pointed out that even if no shortcuts (such as the beam search) are taken, the Viterbi algorithm is not guaranteed to find the most probable word string. So tree and Viterbi searches are both suboptimal in different ways, and the choice between them must be made in accordance with experimental evidence and other circumstances of the task.

6.3 A* Search

This section considers the general problem of tree search. Its applicability is therefore not confined to speech recognition, even though it will be expressed in its terminology. In view of the fact that we have used the symbol **A** to denote a string of acoustic symbols, the term "A* search" [1] [2] may be somewhat confusing. We use it nevertheless, because that

---

2. We pointed out in section 5.5 that the number of language model equivalence classes need not necessarilly determine the size of the search state space, but taking advantage of that observation has its own costs.

designation of the relevant algorithm is so widespread; we simply caution the reader to stay alert and not panic.

Any tree search must be based on some evaluation criterion related to the search's purpose. Formally, the search is for a path $\mathbf{w}_1^n \doteq w_1, w_2, \ldots, w_n$ (representing the complete utterance) that maximizes some stated criterion function $g(\mathbf{w}_1^n)$, where $g(\mathbf{w}_1^k)$ is defined for all lengths $k = 1, 2, \ldots, n$.

Let us define an auxiliary function $d(\mathbf{w}_1^k)$ and with its help yet another function

$$F(\mathbf{w}_1^k) \doteq g(\mathbf{w}_1^k) + d(\mathbf{w}_1^k) \tag{2}$$

Let us further assume that $d(\mathbf{w}_1^n) = 0$, and that

$$d(\mathbf{w}_1^k) \geq g(\mathbf{w}_1^k \| \mathbf{z}_1^{n-k}) - g(\mathbf{w}_1^k) \qquad \text{for all word strings } \mathbf{z}_1^{n-k} \tag{3}$$

where $\mathbf{w}_1^k \| \mathbf{z}_1^{n-k}$ denotes the concatenation of the two sequences involved. Then it follows directly from (2) and (3) that

$$\text{If } F(\mathbf{w}_1^n) \geq F(\tilde{\mathbf{w}}_1^k), \text{then } g(\mathbf{w}_1^n) \geq g(\tilde{\mathbf{w}}_1^k \| \mathbf{z}_1^{n-k}) \text{ for all } \mathbf{z}_1^{n-k} \tag{4}$$

That is, if $F(\mathbf{w}_1^n) \geq F(\tilde{\mathbf{w}}_1^k)$ then no continuation of the path $\tilde{\mathbf{w}}_1^k$ can turn out to be better[3] than the particular complete path $\mathbf{w}_1^n$. If this is the case, therefore, all the continuations of $\tilde{\mathbf{w}}_1^k$ can be completely eliminated as candidates for best path.

As a result of observation (4), the following search algorithm finds the best path:

A* Search [1] [2]
1. Insert into the stack all the single-branch paths corresponding to the words of the vocabulary. Arrange the entries in descending order of $F(v_{i_j})$ (i.e., $F(v_{i_j}) \geq F(v_{i_{j+1}})$), $v_{i_j} \in \mathscr{V}$.
2. Take the top entry $F(\mathbf{w}_1^k)$ off the stack. If $k = n$ then stop—this $\mathbf{w}_1^n$ is the best path—else use $F(\ )$ to evaluate all extensions $\mathbf{w}_1^k \| v_j$ for $j = 1, 2, \ldots, |\mathscr{V}|$ and insert them into the stack according to their value $F(\mathbf{w}_1^k \| v_j)$, maintaining the stack in order of descending value.
3. Repeat above step until the stopping criterion in step 2 is satisfied.

---

3. That is, no continuation can turn out better as measured by the evaluation function $g(\ )$.

6.4 Stack Algorithm for Speech Recognition

The aim of a hypothesis search for speech recognition is to find $\hat{\mathbf{W}}$ satisfying (1). We propose to do so using a sequential search analogous to the A* algorithm. The partial values analogous to $F(\mathbf{w}_1^k)$ of the preceding section might naturally be[4]

$$F(\mathbf{w}_1^k) \doteq \max_{\mathbf{z}_1^r} P(\mathbf{a}_1^m, \mathbf{w}_1^k \| \mathbf{z}_1^r) \tag{5}$$

except that the stopping rule (step 2 in the algorithm presented in section 6.3) will have to be adjusted (we do not know the length n of the utterance) to read something like: "Stop when the top of the stack contains an entry $\mathbf{w}_1^k$ whose value is

$$F(\mathbf{w}_1^k) = P(\mathbf{a}_1^m, \mathbf{w}_1^k)$$

(i.e., the maximum in (5) is attained by the empty word sequence $\mathbf{z}_1^0$) [3]."

To be practical, we will now successively modify the definition of $F(\mathbf{w}_1^k)$. Let

$$l(k) \doteq \arg \max_{0 \le l \le m} P(\mathbf{a}_1^l, \mathbf{w}_1^k) \tag{6}$$

Thus $l(k)$ can be thought of as the length of the initial substring of $\mathbf{a}_1^m$ that is "due" to $\mathbf{w}_1^k$. So our first modification is[5]

$$F(\mathbf{w}_1^k) \doteq \log P(\mathbf{a}_1^{l(k)}, \mathbf{w}_1^k) + \log \left[\max_{\mathbf{z}_1^r} P(\mathbf{a}_{l(k)+1}^m, \mathbf{z}_1^r | \mathbf{w}_1^k) \right] \tag{7}$$

Note that now we really have the additive form $F = g + d$.

The problem is how to evaluate the second term of the right-hand side of (7). To do this exactly is clearly impossible: Too much work at run time or too much storage is required. But the algorithm will work if we overbound d.

One way to proceed is to create an *empirical overbound*. That is, let $d(\mathbf{a}_{l(k)+1}^m, \mathbf{w}_1^k)$ be simply a linear function of $(m - l(k))$, the length of the tail of $\mathbf{a}_1^m$, whose value is related to the logarithm of the maximum value of $P(\mathbf{a}_{l(k)+1}^m, \mathbf{z}_1^r | \mathbf{w}_1^k)$ observed during training. A precise derivation requires

---

4. $\mathbf{z}_1^r$ denotes word strings of length r, and $\mathbf{A} = \mathbf{a}_1^m = a_1, a_2, \ldots, a_m$ denotes the speech data to be recognized.

5. Remember that while "$\|$" denotes concatenation of strings, "$|$" still denotes conditioning in probability functions.

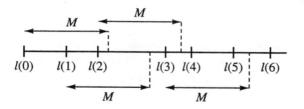

$$i_1 = 3, i_2 = 3, i_3 = 4, i_4 = 6$$

Figure 6.2
Finding word sequence lengths corresponding to at least M acoustic output symbols

some manipulation:

1. After the final training iteration, align intervals in the observed sequence **A** with the words of the script. That is, use your acoustic word models and forced (Viterbi) alignment[6] to find subsequences $\mathbf{a}_{l(j-1)+1}^{l(j)} = a_{l(j-1)+1}, \ldots, a_{l(j)}$ that were "caused" by the words w_j of the training script. As defined in (6), $l(1), l(2), \ldots$ are the endpoints in **A** corresponding to the spoken words $w_1, w_2, \ldots$.

2. Select an observed substring length M. It should account on the average for, say, five to ten words.

3. Find a sequence of integers $i_1 \le i_2 \le \ldots \le i_l \le \ldots$ such that[7]

$$l(i_r - 1) - l(r - 1) < M \text{ and } l(i_r) - l(r - 1) \ge M$$

4. Compute slopes of the per output symbol rise of the logarithms of probabilities

$$\frac{M}{l(i_r) - l(r - 1)} \log P(\mathbf{a}_{l(r-1)+1}^{l(i_r)}, w_r, w_{r+1}, \ldots, w_{i_r} | \mathbf{w}_1^{r-1}), r = 1, 2, \ldots,$$

and save them in a set $\mathscr{C}$.

5. Define $\Delta_{\max}$ and Δ_{av} to be the maximal and average values (slopes) found in the set $\mathscr{C}$. $\Delta_{\max}$ will be regarded as the empirical value of the maximum rise per observed acoustic symbol of the quantity $\log P(\mathbf{a}_{l(k)+1}^m, \mathbf{z}_1^r | \mathbf{w}_1^k)$.

6. See section 3.5.

7. That is, i_r is the least integer such that the word sequence $w_r, w_{r+1}, \ldots, w_{i_r}$ aligns with an observed substring at least as long as M, as figure 6.2 illustrates.

6. In the A* search, use as evaluation function (m is the total length of the observed sequence $\mathbf{A}$)

$$F(\mathbf{w}_1^k) = \log P(\mathbf{a}_1^{l(k)}, \mathbf{w}_1^k) + (m - l(k))\Delta_{\max} \tag{8}$$

6.5 Modifications of the Tree Search

Even though the d function $\Delta_{\max}$ is defined only experimentally, the tree search based on the evaluation function (8) will usually find the maximum likelihood path, albeit slowly. In fact, suppose $\mathbf{w}_1^k$ is the initial string of the correct path, and w_{k+1} is the next correct word. Then the increment in the evaluation function is

$$
\begin{aligned}
F(\mathbf{w}_1^k, w_{k+1}) - F(\mathbf{w}_1^k) = {} & \log P(\mathbf{a}_1^{l(k+1)}, \mathbf{w}_1^{k+1}) - \log P(\mathbf{a}_1^{l(k)}, \mathbf{w}_1^k) \\
& - (l(k+1) - l(k))\Delta_{\max}
\end{aligned} \tag{9}
$$

which is a negative quantity if $\Delta_{\max}$ has its intended value. Therefore, after the correct path floats to the top and is extended, the next entry on top of the stack will too frequently constitute a backtrack.

If instead of $\Delta_{\max}$ we use Δ_{av} then, on the average, the increment (9) will be 0. This will speed up the search at the price of a positive probability of search error. Obviously, there is a compromise in the value of Δ that can be found experimentally. Actually, the best Δ may even be less than Δ_{av}, because for a wrong next word w_{k+1}^* the increment

$$
\begin{aligned}
F(\mathbf{w}_1^k \| w_{k+1}^*) - F(\mathbf{w}_1^k) = {} & \log P(\mathbf{a}_1^{l(k+1)}, \mathbf{w}_1^k \| w_{k+1}^*) - \log P(\mathbf{a}_1^{l(k)}, \mathbf{w}_1^k) \\
& - (l(k+1) - l(k))\Delta_{av}
\end{aligned}
$$

will have a negative expectation. This follows from the fact that the value of Δ_{av} was derived from correct paths, and so the likelihood increment along incorrect paths should on average be less.

6.6 Multiple-Stack Search

The amount of searching can be controlled yet another way based on multiple stacks [5]. The procedure's heuristic justification presented here is due to Doug Paul [4]. Certain preliminary restrictions necessary to make the justification rigorous should not disturb the reader, as they will be removed in practice, as described in section 6.6.3.

6.6.1 First Algorithm

We will assume that the recognizer (a) uses a unigram language model, (b) allows for no phonetic dependence across word boundaries, and (c) makes hard decisions about word boundaries in the string $\mathbf{A}$.[8]

As a result of these restrictions we define the evaluation function[9]

$$g(l, \mathbf{w}_1^k) \doteq \log P(\mathbf{a}_1^l, \mathbf{w}_1^k) = \max_{l_1,\ldots,l_{k-1}} \sum_{i=1}^{k} \log P(\mathbf{a}_{l_{i-1}+1}^{l_i}, w_i) \qquad l_0 = 0, l_k = l \tag{10}$$

Let us further define (both the string length j and the elements of the string $\mathbf{w}_1^j$ are variables)

$$g^*(l) \doteq \max_{\mathbf{w}_1^j} g(l, \mathbf{w}_1^j) \tag{11}$$

and with its help the evaluation function $F^*(l, \mathbf{w}_1^k)$ (indicating how well the word string $\mathbf{w}_1^k$ accounts for the acoustic string $\mathbf{a}_1^l$ compared to the "best" word string that defines (11))

$$F^*(l, \mathbf{w}_1^k) \doteq g(l, \mathbf{w}_1^k) - g^*(l) \tag{12}$$

Clearly,

$$F^*(l, \mathbf{w}_1^k) \leq 0 \tag{13}$$

Our restricted recognizer is searching for the word sequence

$$\hat{\mathbf{W}} = \arg \max_{\mathbf{w}_1^k} F^*(m, \mathbf{w}_1^k) \tag{14}$$

Suppose now that $\hat{\mathbf{W}} = \hat{\mathbf{w}}_1^n$. Then by (10) there exist integers $\hat{l}_0 = 0, \hat{l}_1, \ldots, \hat{l}_{n-1}, \hat{l}_n = m$[10] such that

$$F^*(\hat{l}_k, \hat{\mathbf{w}}_1^k) = 0 \qquad \text{for } k = 0, 1, \ldots, n \tag{15}$$

---

8. That is, when computing the probability of a word string $\mathbf{w}_1^k$ generating an acoustic string $\mathbf{A}$, the recognizer decides for each word $w_i, i \in \{1, 2, \ldots, k\}$ the position in $\mathbf{A}$ of the segment $a_{l_{i-1}}, \ldots, a_{l_i-1}$ generated as a result of w_i's having been spoken.

9. $l_1, l_2, \ldots, l_{k-1}$ are the optimal endpoints for the words $w_1, w_2, \ldots, w_{k-1}$ if conditions (a) and (b) are satisfied. Condition (c) is necessary for the second equality in (10) to hold.

10. We remind the reader that m denotes the length of the observed acoustic string.

In fact, let $\hat{l}_0, \hat{l}_1, \ldots, \hat{l}_{n-1}, \hat{l}_n$ be the speech endpoints of the words of the most probable string $\hat{\mathbf{w}}_1^n$, and suppose that for some value of k, $F^*(\hat{l}_k, \hat{\mathbf{w}}_1^k) < 0$. Then by definition (11) a word sequence $\mathbf{w}_1^r$ would have to exist for which $F^*(\hat{l}_k, \mathbf{w}_1^r) = 0$, and in that case $\mathbf{w}_1^r \| \hat{\mathbf{w}}_{k+1}^n$ would have a higher probability than $\hat{\mathbf{w}}_1^n$, contradicting (14).

By similar reasoning it is clear that if $F^*(l, \mathbf{w}_1^r) < 0$ for some $\mathbf{w}_1^r$, then no continuation of $\mathbf{w}_1^r$ whose speech boundary takes place at time l can result in the most probable word string. In fact, because of (11) and (12) there must exist a word string $\tilde{\mathbf{w}}_1^s$ for which $F^*(l, \tilde{\mathbf{w}}_1^s) = 0$, so that any continuation of $\tilde{\mathbf{w}}_1^s$ results in a more probable sequence than the same continuation of $\mathbf{w}_1^r$.

It thus follows from (13), (15) and the last conclusion that under the imposed restrictions (a) through (c), the following algorithm will find the most probable word sequence:

1. For each word v and $l = 1, 2, \ldots, m$, compute $F^*(l, v)$.
2. For any v such that $F^*(1, v) = 0$, compute the extensions $F^*(1 + i, \mathbf{w}_1^2 = v \| v')$ for all words v' and $i = 1, 2, \ldots, m - 1$.
3. Do $l = 2$ to $m - 1$:
(a) For any $\mathbf{w}_1^k$ such that $F^*(l, \mathbf{w}_1^k) = 0$, compute the extensions $F^*(l + i, \mathbf{w}_1^{k+1} = \mathbf{w}_1^k \| v')$ for all words v' and $i = 1, 2, \ldots, m - l$.
(b) End
4. A sequence $\hat{\mathbf{w}}_1^n$ must exist such that $F^*(m, \hat{\mathbf{w}}_1^n) = 0$. It is the best sequence, and the algorithm will find it.

6.6.2 A Multistack Algorithm

It is obvious that as it stands, the above algorithm cannot be carried out because there is no way (save an exhaustive search) to compute the maxima $g^*(l)$. Fortunately it turns out that we do not need them.

Suppose that for each $l = 1, \ldots, m$ we establish a stack into which we will insert in descending order the values $g(l, \mathbf{w}_1^k)$ for those sequences $\mathbf{w}_1^k$ previously examined [5]. At any given time t of the stack development, we will define

$$g_t^*(l) \doteq \max_{\substack{\text{entries at time } t}} g(l, \mathbf{w}_1^k) \tag{16}$$

and correspondingly

$$F_t^*(l, \mathbf{w}_1^k) \doteq g(l, \mathbf{w}_1^k) - g_t^*(l) \tag{17}$$

It is clear that for the best (most probable) sequence $\hat{\mathbf{w}}_1^n$ there will exist integers $\hat{l}_k$, $k = 1, \ldots, n$ (see the discussion leading to (15)) such that

$F_t^*(\hat{l}_k, \hat{\mathbf{w}}_1^k) = 0$ at all times t subsequent to the instant at which the entry $g(\hat{l}_k, \hat{\mathbf{w}}_1^k)$ is inserted into the $\hat{l}_k^{th}$ stack. The following algorithm will assure that such entries will indeed be created:

1. For each word v and $l = 1, 2, \ldots, m$, compute $g(l, v)$ and insert them into their proper position in the lth stacks.
2. Establish $g_1^*(1)$.
3. Create all extensions $g(1 + i, \mathbf{w}_1^2 = v\|v'), i = 1, 2, \ldots, m - 1$ of all words v in stack 1 for which $F_1^*(1, v) = 0$, and insert them into their stacks.[11]
4. Do $l = 2$ to $m - 1$:
(a) Establish $g_l^*(l)$.
(b) Create all extensions $g(l + i, \mathbf{w}_1^{k+1} = \mathbf{w}_1^k\|v')$, $i = 1, \ldots, m - 1$ of all sequences $\mathbf{w}_1^k$ in stack l for which $F_l^*(l, \mathbf{w}_1^k) = 0$, and insert them into their stacks.
(c) End
5. The top entry of stack m is the desired best word sequence.

The reason the above algorithm works is that (a) if we are assured that by the time the lth stack is extended it already contains the best word sequence that could ever be placed into it[12], then we need extend only those sequences for which $F_l^*(l, \mathbf{w}_1^k) = 0$, and (b) we actually extend the lth stack only after it contains all the sequences that will ever be inserted into it.

Obviously, various shortcuts exist limiting the necessary computation. For instance, $g(l, \mathbf{w}_1^k)$ needs to be computed only for durations l that might conceivably result from uttering $\mathbf{w}_1^k$.

6.6.3 Actual Multistack Algorithm

It is necessary to stress that the guarantee that the last algorithm will find the best word sequence depends completely on the restrictions (a), (b), and (c) stated at the beginning of Section 6.6.1. Should these restrictions be violated, it would no longer be necessarily true that for the sub-sequences $\hat{\mathbf{w}}_1^k$ of the best sequence $\hat{\mathbf{w}}_1^n$,

$$g(l, \hat{\mathbf{w}}_1^k) = \max_{\mathbf{w}_1'} g(l, \mathbf{w}_1')$$

would hold for any l whatever!

---

11. That is, entry $g(1 + i, \mathbf{w}_1^2 = v\|v')$ is placed into the $(1 + i)^{th}$ stack.
12. That is, whose estimated ending time is l.

Nevertheless, a relatively slight modification of the algorithm of section 6.6.2 will, not surprisingly, lead to a practical procedure that will very rarely fail to find the most probable word string. We must first modify some of the definitions (10), (11), and (17):

$$g^+(l, \mathbf{w}_1^k) \doteq \log P(\mathbf{a}_1^l, \mathbf{w}_1^k) \tag{18}$$

$$g_t^+(l) \doteq \max_{\text{entries at time } t} g^+(l, \mathbf{w}_1^k) \tag{19}$$

$$F_t^+(l, \mathbf{w}_1^k) \doteq g^+(l, \mathbf{w}_1^k) - g_t^+(l) \tag{20}$$

The algorithm now becomes:

1. For each word v and $l = 1, 2, \ldots, m$ compute $g^+(l, v)$ and insert it into its proper position in the lth stack.
2. Establish $g_1^+(1)$.
3. Create all extensions $g^+(1 + i, \mathbf{w}_1^2 = v\|v'), i = 1, 2, \ldots, m - 1$ of all words v in stack 1 for which $F_1^+(1, v) \geq -\tau$, and insert them into their stacks.
4. Do $l = 2$ to $m - 1$:
(a) Establish $g_l^+(l)$.
(b) Create all extensions $g^+(l + i, \mathbf{w}_1^{k+1} = \mathbf{w}_1^k\|v'), i = 1, \ldots, m - l$ of all sequences $\mathbf{w}_1^k$ in stack l for which $F_l^+(l, \mathbf{w}_1^k) \geq -\tau$, and insert them into their stacks.
(c) End.
5. The recognizer decides for the word sequence found as the top entry of stack m.

An appropriate (positive) value of the threshold τ must be experimentally established.

6.7 Fast Match

We have not yet considered a crucial problem encountered in a tree search (and, indeed, a trellis search as well) for a most likely large vocabulary utterance. Namely, if the vocabulary is of size $|\mathscr{V}|$, an extension of a path implies the calculation of $|\mathscr{V}|$ probabilities $P(\mathbf{a}_1^{l(k+1)}, \mathbf{w}_1^k\|v)$, one for each word v of the vocabulary $\mathscr{V}$. For vocabulary sizes in the tens of thousands, this is certainly intolerable.

We need to find a way to extend a path $\mathbf{w}_1^k$ by only those words v which bear some acoustic similarity to the observed sequence $a_{l(k)+1}, a_{l(k)+2}, \cdots$ The task of the *fast match* [6] is to find a limited list of such words. The basic idea involves two steps:

1. Using an easy to evaluate, but perhaps approximate, match measure $Q(v|\mathbf{A})$, create a list of N ($\ll |\mathscr{V}|$, the size of the vocabulary) words whose acoustic realization is expected to be similar to $a_{l(k)+1}, a_{l(k)+2}, \ldots$.
2. For any particular history $\mathbf{w}_1^k$ under consideration, retain in that list only those words v whose probability estimate $Q(v|\mathbf{A})P(v|\mathbf{w}_1^k)$ is among the highest.

Step 1 is referred to as the *acoustic fast match*. We will sketch it in two installments. Our purpose, as always, is to impart the basic idea. Its precise application must be worked out for each case at hand.

Let a word HMM be made up of a concatenation of HMMs corresponding to the phones forming its base form.[13] We can arrange the vocabulary in a tree whose branches are labeled by phones and whose leaves are the words of the vocabulary. The sequence of labels of branches forming the path to the leaf will constitute the base form of the word identified with the leaf and will thus specify the word's HMM. The probability associated with any tree branch will be made proportional to the sum of the (unigram) probabilities of words corresponding to those leaves that are reachable by taking that branch. The tree of figure 6.3 illustrates this for a small vocabulary.

We can search the tree using the following algorithm whose evaluation function $F(\varphi_1^k)$ can be based on the principles discussed in sections 6.4 and 6.5. (We assume that the base form of word v is specified by $\Phi(v) = \varphi_1, \varphi_2, \ldots, \varphi_{n(v)}$.) Naturally, the crucial component of $F(\varphi_1^k)$ will be the probability of generating the appropriate substring $a_{l(k)+1}, a_{l(k)+2}, \ldots$ of the observed acoustic string $\mathbf{A}$ by the composite HMM that results from the concatenation of the HMMs corresponding to the individual phones of φ_1^k.

Acoustic Fast Match
1. Insert into the stack all the single-branch paths going out from the root node of the phonetic tree. Arrange the entries in descending order of $F(\varphi_{i_j})$ (i.e., $F(\varphi_{i_j}) \geq F(\varphi_{i_{j+1}})$).
2. Take the top entry $F(\varphi_1^k)$ off the stack.
(a) If φ_1^k is a complete word ($\varphi_1^k = \Phi(v)$ for some word v), place v on the acoustic fast match list. If this is the Nth word on the list then stop, else continue.

---

13. Thus we are using here the phonetic models described in section 3.2 as the basic building blocks of the word HMM. This is what we wish to use to build our fast match, regardless of what building blocks the recognizer's acoustic model uses (for instance, the fenonic models of chapter 3 or the triphones of chapter 12).

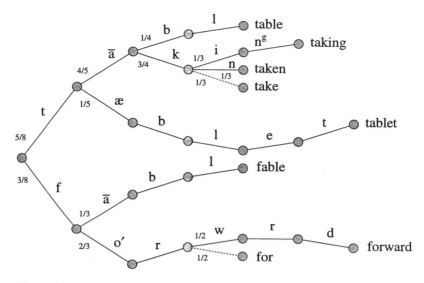

Figure 6.3
Phonetic tree corresponding to a small vocabulary, with branch probabilities
calculated from unigram word probabilities

(b) Else if φ_1^k is not a complete word, evaluate all its extensions $\varphi_1^k\|\varphi$ for
phones φ that can follow φ_1^k in the phonetic tree, and insert these exten-
sions into the stack according to their value $F(\varphi_1^k\|\varphi)$, maintaining the
stack in descending value order.
3. Repeat step 2 until the stopping criterion is satisfied.

Obviously, this procedure will yield N candidate words for a possible
detailed evaluation later during the word hypothesis search. Because the
vast majority of the branches of the phonetic tree will be pruned (that is,
never evaluated), this setup will accelerate the search, but not sufficiently—
additional shortcuts will be needed.

Consider the standard phonetic HMM shown in figure 6.4. It has twelve
transitions t_i of which the first nine produce outputs, and the remaining
three do not. The HMM is specified by its topology, by the transition
probabilities $p(t_i)$, and by the output probabilities $q(a|t_i)$.

Calculating the probability that an initial substring of a string
$a_1, a_2, \ldots, a_j, \ldots$ was produced by the phonetic HMM (which is what
needs to be calculated in each extension step of the acoustic fast match
algorithm) is (relatively) computing intensive. Figure 6.5 shows the top-

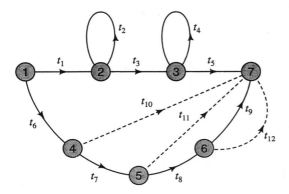

Figure 6.4
Structure of a standard phonetic HMM

ology of the trellis involved. The required probability is equal to the sum $\sum_{s^*} \alpha_i(s^*)$ of probabilities (defined in equation (8) of chapter 2) associated with the last states s^* of each trellis stage. (Note that the trellis transition structure does not even start to repeat before the fifth stage.)

On the other hand, consider the HMM of figure 6.6. Its trellis structure is extremely simple (figure 6.7). In fact, the probability of that HMM's generating an initial substring of a string $a_1, a_2, \ldots, a_j, \ldots$ is given by the formula

$$p(t_1)[q(a_1|t_1) + p(t_2)q(a_1|t_2)[q(a_2|t_1) + p(t_2)q(a_2|t_2)[q(a_3|t_1)$$

$$+ p(t_2)q(a_3|t_2)[q(a_4|t_2) + p(t_2)q(a_4|t_2)[q(a_5|t_2) + \cdots]]]]$$

which can be further simplified if $q(a|t_1) = q(a|t_2) = q(a)$, and $p(t_2) = p$, $p(t_1) = 1 - p$ to result in:

$$(1 - p)q(a_1)[1 + p\,q(a_2)[1 + p\,q(a_3)[1 + p\,q(a_4)[1 + p\,q(a_5)[1 + \cdots]]]]]$$

$$= (1 - p) \sum_{k \geq 1} p^{k-1} \prod_{i=1}^{k} q(a_i)$$

So computation would be considerably accelerated if we could some-how change the HMM structure from that of figure 6.4 into that of figure 6.6. This we now proceed to do.

Remember that the whole idea of the fast match is to eliminate all words v which have little chance of being the causes of the appropriate substring of the observed acoustic string **A**. So if we perform the match of

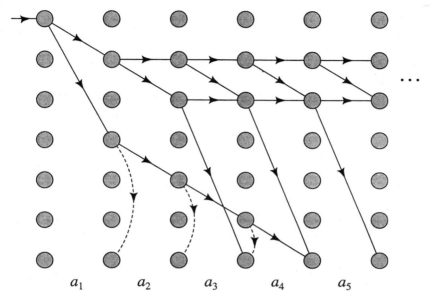

a_1 a_2 a_3 a_4 a_5

Figure 6.5
Topology of trellis corresponding to HMM of figure 6.4

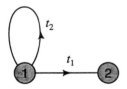

Figure 6.6
Very simple HMM

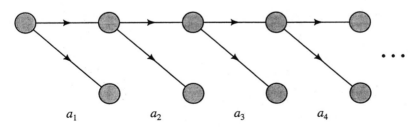

a_1 a_2 a_3 a_4

Figure 6.7
Topology of trellis corresponding to HMM of figure 6.6

the acoustic string with the model of word v and then upper bound its value, and if the upper bound obtained is so small (compared to some fixed threshold) as to eliminate this word v from consideration, then v would surely have been eliminated (by comparison with the same threshold) even if we had used its true match value.

One way to upper bound the match probability is to upper bound all the HMM's output distributions. This can be done by replacing them with the pseudodistribution

$$q(a) = \max_{1 \le i \le 9} q(a|t_i) \tag{21}$$

($q(a)$ is a pseudodistribution because $\sum_a q(a) \ge 1$).

Once the same pseudodistribution $q(a)$ is attached to all the output-producing branches of the HMM of figure 6.4, we can change its topology into that of figure 6.6. If we determine appropriately the transition probability $p(t_1)$ for figure 6.6, then a match of the acoustic string based on the resulting HMM will upper bound approximately the match based on the original HMM of figure 6.4.

A reasonable way to select $p(t_2) = p$ is to determine it so that the expected number of outputs $\bar{L}$ generated by the two HMMs (the original one of figure 6.4 and its "simplified" version in figure 6.6), $a_{l(k)+1}, a_{l(k)+2}$, ..., are the same. $\bar{L}$ must first be calculated for the phonetic HMM corresponding to the phone in question, and then p is taken as the solution of the equation

$$\bar{L} = (1 - p) \sum_{k=1}^{\infty} k p^{k-1} = \frac{1}{1 - p}$$

The acoustic fast match performed on the basis of the simplified phone HMMs (which will determine the evaluation function $F(\varphi_1^k)$ used in the pruning of the phone-based hypothesis tree) then yields a list of N words and their approximate match values $Q(v|\mathbf{A})$. The fast match is completed by retaining only those words from the list whose combined values $Q(v|\mathbf{A})P(v|\mathbf{w}_1^k)$ satisfy

$$\log Q(v|\mathbf{A})P(v|\mathbf{w}_1^k) \ge \max_{\tilde{v}} \log Q(\tilde{v}|\mathbf{A})P(\tilde{v}|\mathbf{w}_1^k) - \tau$$

for some appropriately chosen threshold value τ.

One last remark: We motivated the use of the pseudodistribution (21) by the comparison of the resulting match values with a fixed threshold. However, the acoustic fast match described in this section does not involve any fixed thresholds. Rather, it works via a comparison of matches with each other. So tree paths $\tilde{\varphi}_1^l$ will be eliminated from the list because the upper-bound evaluations $F(\varphi_1^k)$ of other paths will be higher than the upper bound $F(\tilde{\varphi}_1^l)$. Thus we cannot actually guarantee that the eliminated paths would have been purged had we not resorted to upper bounding. Although the motivation that we gave reflects accurately what was in the mind of the researchers who originated the idea, the true justification of the method lies in its good performance in practice.

6.8   The Cost of Search Shortcuts

In the preceding chapter we have used the Viterbi algorithm to search for the word string $\hat{\mathbf{W}}$ that the recognizer will put out. The search shortcuts of that method (which itself is a shortcut) are the beam search, multipass searches,[14] and differing search and language model state spaces (sections 5.3, 5.4, and 5.5, respectively). The multistack algorithm of the present chapter (section 6.6.3) is also suboptimal,[15] and its practical search capabilities are further weakened by the fast match (section 6.7).

In either of these cases, how can we assess the damage of the shortcuts? In any test situation where the "truth" is specified (by comparison of the found word string with the human transcription of the speech), we know whether an error was committed. But we don't know what it was due to: bad models, irregular and/or noise-contaminated speech, or to the shortcuts themselves. The way to estimate the damage is very straightforward:

Let $\hat{\mathbf{W}}$ be the recognizer output string, let $\mathbf{W}^*$ be the truth, and suppose that $\hat{\mathbf{W}} \neq \mathbf{W}^*$. Then the recognizer will surely have enough computing power to compare $P(\mathbf{A}|\hat{\mathbf{W}})P(\hat{\mathbf{W}})$ with $P(\mathbf{A}|\mathbf{W}^*)P(\mathbf{W}^*)$. If the first product is larger, then errors are not due to search shortcuts.[16] If the second product is larger, then a search error has been committed.

We can of course refine this test to find more about the causes of failure. For instance, in the stack search case we can find out whether the

14. Allowing for more complex than bigram language models.

15. Its A^* nature having been compromised by the elimination of the three conditions listed at the start of section 6.6.1 and by the consequent introduction of the threshold $-\tau$ into the algorithm.

16. This does not necessarily mean that $\hat{\mathbf{W}}$ is the best word string obtainable with the current models, only that the truth would have been unobtainable even without shortcuts.

correct path was eliminated by the fast match: that is, whether a particular erroneous word in $\hat{\mathbf{W}}$ failed to appear on the list supplied by the fast match.

Or, if the threshold value τ is suspected to be the culprit, we can confirm or reject it as the cause. Let $P(\mathbf{A}|\hat{\mathbf{W}})P(\hat{\mathbf{W}}) < P(\mathbf{A}|\mathbf{W}^*)P(\mathbf{W}^*)$ and let k be the lowest index such that $w_{k+1}^* \neq \hat{w}_{k+1}$. All we need to do is to preserve the final values $g_T^+(l)$ and check whether $\log P(\mathbf{a}_1^l, \mathbf{w}_1^k) < g_T^+(l) - \tau$ for all $l = 1, 2, \ldots, m$. If so, then the search error is due to the threshold value.

Similarly, in Viterbi search, in cases where $P(\mathbf{A}|\hat{\mathbf{W}})P(\hat{\mathbf{W}}) < P(\mathbf{A}|\mathbf{W}^*)P(\mathbf{W}^*)$, we can see whether the most probable path through the composite HMM of $\mathbf{W}^*$ has a probability which exceeds that for $\hat{\mathbf{W}}$. If so, then the error was due not to the Viterbi procedure itself, but to one of its shortcuts (e.g., beam search or multipass).

6.9 Additional Reading

The method of Bahl et al. [5] [7] actually differs from that of [4] by more than the use of multiple stacks. It incorporates an envelope (obtained by backtracing) that considerably speeds up the search.

Reference [8] discusses aspects of A^* algorithms. Some of the authors introduce very innovative bidirectional search strategies having desirable computational efficiency [9]. Lennig and coworkers present yet another alternative [10].

As pointed out in section 6.7, large vocabularies require that the search be accelerated by a fast match. Reference [11] explores further the one discussed in section 6.7. Yet another fast match is introduced in [12]. Finally, a totally different idea based on majority decisions involving elements of observed discrete acoustic output strings leads to the polling fast match [13].

References

[1] P.E. Hart, N.J. Nilsson, and B. Raphael, "A formal basis for the heuristic determination of minimum cost paths," *IEEE Transactions on Systems Science and Cybernetics*, SSC-4, no. 2, pp. 100–07, 1968.

[2] P.E. Hart, N.J. Nilsson, and B. Raphael, Correction to "A formal basis for the heuristic determination of minimum cost paths," *SIGART Newsletter*, no. 37, pp. 28–29, 1972.

[3] F. Jelinek, "A fast sequential decoding algorithm using a stack," *IBM Journal of Research Development*, vol. 13, pp. 675–85, Nov. 1969.

[4] D.B. Paul, "An essential A* stack decoder algorithm for continuous speech recognition with a stochastic language model," *Proceedings of the 1992 International Conference on Acoustics, Speech, and Signal Processing*, San Francisco, CA, March 1992.

[5] L.R. Bahl, P.S. Gopalakrishnan, and R.L. Mercer, "Search issues in large vocabulary speech recognition," *Proceedings of the 1993 IEEE Workshop on Automatic Speech Recognition*, Snowbird, UT, 1993.

[6] L.R. Bahl, S.V. De Gennaro, P.S. Gopalakrishnan, R.L. Mercer, "A fast approximate acoustic match for large vocabulary speech recognition," *IEEE Transactions on Speech and Audio Processing vol. 1, no. 1, pp. 59–67, January 1993.*

[7] P.S. Gopalakrishnan, L.R. Bahl, and R.L. Mercer, "A tree search strategy for large-vocabulary continuous speech recognition," *Proceedings of the 1995 International Conference on Acoustics, Speech, and Signal Processing*, vol. I, pp. 572–75, Detroit, MI, May 1995.

[8] P. Kenny, R. Hollan, V.N. Gupta, M. Lenning, P. Mermelstein, an D. O'Shaughnessy, "A*-admissible heuristics for rapid lexical access," *IEEE Transactions on Speech and Audio Processing*, vol. 1, no. 1, pp. 49–58, January 1993.

[9] Z. Li, G. Boulianne, P. Laboute, M. Barszcz, H. Garudadri, and P. Kenny, "Bi-directional graph search strategies for speech recognition," *Computer Speech and language*, vol. 10, no. 4, pp. 295–321, October 1996.

[10] V.N. Gupta, M. Lennig, and P. Mermelstein, "Fast search strategy in a large vocabulary word recognizer," *Journal of the Acoustical Society of America*, vol. 84, December 1988.

[11] P.S. Gopalakrishnan and L.R. Bahl, "Fast matching techniques," in *Automatic Speech and Speaker Recognition*, C-H. Lee, F.K. Soong, and K.K. Paliwal, eds., pp. 413–28, Klewer Academic Publishers, Norwell, MA, 1996.

[12] P. Kenny, P. Labute, Z. Li, R. Hollan, M. Lennig, and D. O'Shaughnessy, "A new fast match for very large vocabulary continuous speech recognition," *Proceedings of the 1993 International Conference on Acoustics, Speech, and Signal Processing*, vol. II, pp. 656–59, Minneapolis, MN, April 1993.

[13] L.R. Bahl, R. Bakis, P.V. deSouza, and R.L. Mercer, "Obtaining candidate words by polling in a large vocabulary speech recognition system," *Proceedings of the 1988 International Conference on Acoustics, Speech, and Signal Processing*, vol. I, pp. 489–92, New York, April 1988.

Chapter 7

Elements of Information Theory

7.1 Introduction

In speech recognition, concepts originating in information theory [1] [2] are used in two ways: to measure the difficulty of tasks and to provide a criterion for running various training algorithms. In this chapter we will introduce these concepts more or less heuristically, but we will provide a smattering of theorems to give a firm basis to our hand-waving. In the next chapter we will use information theory to estimate the relative difficulty of recognition tasks and the quality of language models. Following chapter 9, information measures will provide a helpful selection criterion in the development of language and phonetic models.

This chapter introduces the following concepts:

1. Entropy as a measure of information content or disorder:

- As a function of the underlying probability distribution,

$$H(p_0, p_1, \ldots, p_{L-1}) = -\sum_{i=0}^{L-1} p_i \log p_i$$

- As a function of a random variable X,

$$H(X) = -\mathbf{E}[\log P\{X\}]$$

2. Conditional entropy,

$$H(X|Y) = -\mathbf{E}[\log P\{X|Y\}]$$

3. Average mutual information,

$$I(X; Y) = -\mathbf{E}\left[\log \frac{P\{X, Y\}}{P\{X\}P\{Y\}}\right]$$

4. Divergence,

$$D(\mathbf{P}\|\mathbf{Q}) = \mathbf{E}_P\left[\log \frac{P\{X\}}{Q\{X\}}\right]$$

Readers familiar with these concepts and their properties are invited to skip this chapter entirely. Here is the chapter's outline for those who are less sophisticated:

In section 7.2 we derive the functional form of $H(p_0, p_1, \ldots, p_{L-1})$ as a consequence of four properties that an intuitively acceptable information measure should possess. In section 7.3 we use this form to deduce additional properties of entropy. In section 7.4 we introduce entropy and conditional entropy as functions of random variables and deduce their properties. Section 7.5 is dedicated to the proof of Shannon's noiseless source coding theorem, which leads directly to the functional form of entropy[1] and allows an operational interpretation of that measure. In section 7.7 we use heuristics to derive the functional form of a mutual information measure and state Shannon's coding theorem for transmission of information through noisy channels, which provides the operational justification for the mutual information function.

7.2   Functional Form of the Basic Information Measure

Let us first derive a measure of information by listing some properties that it should possess.[2] The measure will be statistical, and it will apply, to start with, to the simplest of information sources. For our purposes, the simplest information source is a device that generates an output symbol x from a finite set $\mathcal{X} = \{0, 1, \ldots, M-1\}$ of possibilities. The source is characterized by its output probability distribution $P(x)$, and it chooses its outputs independently of all previous choices (see figure 7.1). Thus

$$P(x_1, x_2, \ldots, x_n) = \prod_{i=1}^{n} P(x_i) \qquad x_i \in \mathcal{X}$$

---

1. The same form as that derived earlier in section 7.2.

2. The measure of information will have a unique form (up to a multiplicative constant). It can be derived either axiomatically, as we are about to do, or operationally. In the latter case it falls out from a coding theorem, as will be seen in section 7.5.

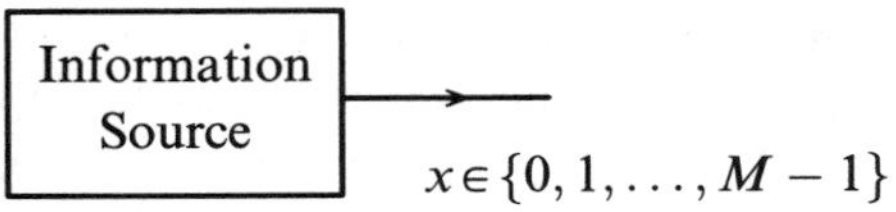

```
┌─────────────┐
│ Information │ ──────────▶
│   Source    │ ──────────
└─────────────┘     $x \in \{0, 1, \ldots, M-1\}$
```

$$P\{x_1, x_2, \ldots, x_n\} = \prod_{i=1}^{n} P(x_i)$$

Figure 7.1
A discrete memoryless information source

As the source provides information to an observer, it can be regarded as removing uncertainty from the observer's "mind" about what the output is going to be. It is clear that the source's *information capacity* (the average amount of information that it provides with each of its outputs) should be a function of the output probability distribution $p_x = P(x), x \in \mathcal{X}$.

Let us then denote by $H(p_0, p_1, \ldots, p_{M-1})$ the information measure we are about to develop. We want to derive its functional form. We will do so step by step, by successively introducing functional properties that it must possess,[3] and then deriving their consequences. Only four properties will be needed.

Let

$$f(M) \doteq H\left(\frac{1}{M}, \frac{1}{M}, \ldots, \frac{1}{M}\right) \tag{1}$$

denote the value of the measure for the uniform distribution over an output alphabet of size M.

Assume for a moment that the alphabet consists of $M = NL$ uniformly distributed symbols.

1. An information measure worthy of its name should have a functional form satisfying the equation

$$f(NL) = f(N) + f(L) \tag{2}$$

Relation (2) holds because we can group the NL symbols into N groups of L symbols and provide the information about any output symbol in two steps: by first identifying the group to which it belongs (and there are

---

3. It is our expectation that these properties will seem so natural that nobody could accept an information measure that would violate any of them.

N equally probable groups, so this step has information value $f(N)$), and then by identifying the particular symbol in question from the group already identified (and since there are L equally probable symbols in the group, the worth of this identification is $f(L)$).

In follows immediately by repeated application of equation (2) that

$$f(N^k) = f(N) + f(N^{k-1}) = f(N) + (f(N) + f(N^{k-2})) = \cdots = kf(N)$$
(3)

2. Since specifying an output symbol from a uniformly distributed alphabet of size $M + 1$ must provide more information (must remove more uncertainty) than specifying an output from an alphabet of size M, then

The value of the function $f(M)$ must increase monotonically with M.   (4)

Now given $M \geq 2$, for any fixed positive integer r there exists an integer k such that

$$M^k \leq 2^r < M^{k+1}$$
(5)

As a consequence of (3), (4), and (5),

$$kf(M) = f(M^k) \leq f(2^r) = rf(2) < f(M^{k+1}) = (k+1)f(M)$$

Dividing through by $rf(M)$ we get

$$\frac{k}{r} \leq \frac{f(2)}{f(M)} < \frac{k+1}{r}$$
(6)

Returning to (5) and taking logarithms of its three terms, we get

$$k \log M \leq r \log 2 < (k+1) \log M$$

and as a consequence

$$\frac{k}{r} \leq \frac{\log 2}{\log M} < \frac{k+1}{r}$$
(7)

Therefore, taking into account both the right inequality of (7) and the left inequality of (6), we get

$$\frac{\log 2}{\log M} - \frac{f(2)}{f(M)} < \frac{k+1}{r} - \frac{f(2)}{f(M)} \leq \frac{k+1}{r} - \frac{k}{r} = \frac{1}{r}$$

Working next with the left inequality of (7) and the right inequality of (6), we get

$$\frac{\log 2}{\log M} - \frac{f(2)}{f(M)} \geq \frac{k}{r} - \frac{f(2)}{f(M)} > \frac{k}{r} - \frac{k+1}{r} = -\frac{1}{r}$$

Consequently,

$$\left| \frac{\log 2}{\log M} - \frac{f(2)}{f(M)} \right| < \frac{1}{r} \tag{8}$$

Since r was arbitrary, we can let it increase without limit and thus conclude that

$$f(M) = \frac{f(2)}{\log 2} \log M \tag{9}$$

3. An information measure must possess another property, as follows. Let q_i and t_i be nonnegative numbers such that $\sum_{i=1}^{k} q_i + \sum_{i=1}^{n} t_i = 1$, and define $p \doteq q_1 + q_2 + \cdots + q_k$ and $(1-p) \doteq t_1 + t_2 + \cdots + t_n$. Then the $H(\ )$ function should satisfy[4]

$$H(q_1, q_2, \ldots, q_k, t_1, t_2, \ldots, t_n) = H(p, 1-p) + pH\left(\frac{q_1}{p}, \frac{q_2}{p}, \ldots, \frac{q_k}{p}\right)$$

$$+ (1-p)H\left(\frac{t_1}{1-p}, \frac{t_2}{1-p}, \ldots, \frac{t_n}{1-p}\right) \tag{10}$$

Equation (10) says that the output of a source governed by the distribution $q_1, q_2, \ldots, q_k, t_1, t_2, \ldots, t_n$ can be specified in two steps: First, the observer will be told whether the output belonged to the first k or to the last n symbols, and this information is worth $H(p, 1-p)$ units. Then, if the output belonged to the first k symbols, which happens with probability p, the actual symbol will be identified, and thus $H(q_1/p, q_2/p, \ldots, q_k/p)$ units of information will be supplied; or, if the output belonged to the last n symbols, which happens with probability $(1-p)$, the actual symbol will be identified and thus $H(t_1/(1-p), t_2/(1-p), \ldots, t_n/(1-p))$ units of information will be supplied.

Consider now the consequences of formula (10). Select any rational value for the probability p, that is, $0 < p = r/s < 1$. Then in (10) we can let $k = r$, $n = s - r$, and $q_i = t_j = 1/s$ for all i and j. Substituting into (10) and rearranging, we then get

---

4. Equations (2) and (10) are closely related. In fact, for $N = 2$, and L arbitrary, (2) is a consequence of (10).

$$H\left(\frac{r}{s},\frac{s-r}{s}\right) = H\left(\frac{1}{s},\ldots,\frac{1}{s},\frac{1}{s},\ldots,\frac{1}{s}\right) - \frac{r}{s}H\left(\frac{1}{r},\ldots,\frac{1}{r}\right)$$
$$- \frac{s-r}{s}H\left(\frac{1}{s-r},\ldots,\frac{1}{s-r}\right)$$

or, using for $p = \dfrac{r}{s}$ the definition (1) and the result (9),

$$
\begin{aligned}
H(p,1-p) &= f(s) - pf(r) - (1-p)f(s-r) \\
&= \frac{f(2)}{\log 2}[\log s - p\log r - (1-p)\log(s-r)] \\
&= \frac{f(2)}{\log 2}\left[-p\log\frac{r}{s} - (1-p)\log\frac{s-r}{s}\right] \\
&= \frac{f(2)}{\log 2}[-p\log p - (1-p)\log(1-p)]
\end{aligned}
\tag{11}
$$

For rational values of p, the relation (11) specifies $H(p, 1-p)$ up to a constant determined by our choice of the value of $f(2)$, the amount of information provided in deciding between two equiprobable alternatives (see (1)). We might as well choose arbitrarily

$$f(2) = \log 2 \tag{12}$$

so that, for rational p we get

$$H(p,1-p) = -p\log p - (1-p)\log(1-p) \tag{13}$$

Using definition (1) and (13) for $p = \frac{1}{2}$,

$$f(2) = H\left(\frac{1}{2},\frac{1}{2}\right) = \left[\frac{1}{2}\log 2 + \frac{1}{2}\log 2\right] = \log 2$$

which is consistent with (12).

4. Finally, we will surely want an information measure to have the property

$$H(p,1-p) \text{ is a continuous function of } p. \tag{14}$$

This makes (13) valid for all real $p \in [0,1]$.

We are now ready to derive the final form for $H(p_0,p_1,\ldots,p_{M-1})$. Using formula (10) (letting $k = 1$, $n = M - 1$, and setting $q_1 = p_0$, and $t_i = p_i$, $i = 1, 2, \ldots, M - 1$) we get

$$H(p_0,p_1,\ldots,p_{M-1})$$
$$= H(p_0, 1 - p_0) + p_0 H(1) + (1 - p_0)H\left(\frac{p_1}{1-p_0},\ldots,\frac{p_{M-1}}{1-p_0}\right) \tag{15}$$

So if, analogously to (34), we hypothesize

$$H(p_0, p_1, \ldots, p_{L-1}) = -\sum_{i=0}^{L-1} p_i \log p_i \qquad (16)$$

to be valid for $L = 2, 3, \ldots, M - 1$ (which by (11) and (14) it certainly is for $L = 2$) then, since $H(1) = 0,$[5] we get by equations (11) and (15)

$$H(p_0, p_1, \ldots, p_{M-1})$$

$$= -p_0 \log p_0 - (1 - p_0) \log(1 - p_0) - (1 - p_0) \left[\sum_{i=1}^{M-1} \frac{p_i}{1 - p_0} \log \frac{p_i}{1 - p_0} \right]$$

$$= -\sum_{i=0}^{M-1} p_i \log p_i$$

so that (16) is valid for $L = M$ as well and therefore by induction for all values of L.

We have not yet chosen the base of the logarithm in (16). It is usually set to be 2 so that $f(2) = \log 2 = 1$. In this case, information content is measured in *bits*. Note that the information content of a uniformly distributed binary source (its observer deciding between two equally likely alternatives) is then equal to one bit.

Thus from the four "logical" assumptions (2), (4), (10), and (14) we have now derived the unique formula[6] for an information measure, (16). The functional form (16) is known to physics under the name *entropy*, and it is thought to measure disorder.

7.3   Some Mathematical Properties of Entropy

We will now derive further properties of the entropy function and point out their intuitive interpretations. In the previous section we listed the following three properties as axioms:

5. By equation (1), $H(1) = f(1)$, and by equation (3), $f(1) = kf(1)$ for all k, and therefore $f(1) = 0$.

6. That is, unique up to a multiplicative constant. We remind the reader that we arbitrarily fixed the scale by making the selection (12) and by choosing the base of the logarithm.

(1)

$$H\left(\frac{1}{ML},\frac{1}{ML},\ldots,\frac{1}{ML}\right) = H\left(\frac{1}{M},\frac{1}{M},\ldots,\frac{1}{M}\right) + H\left(\frac{1}{L},\frac{1}{L},\ldots,\frac{1}{L}\right)$$

(2)

$$H\left(\frac{1}{M},\frac{1}{M},\ldots,\frac{1}{M}\right) \text{ is a monotonically increasing function of } M$$

(3)

$$H(q_1,q_2,\ldots,q_k,t_1,t_2,\ldots,t_n) = H(p,1-p) + pH\left(\frac{q_1}{p},\frac{q_2}{p},\ldots,\frac{q_k}{p}\right)$$
$$+ (1-p)H\left(\frac{t_1}{1-p},\frac{t_2}{1-p},\ldots,\frac{t_n}{1-p}\right)$$

The following properties are all based on the inequality $\ln x \leq x - 1$. To make the proofs cleaner we will switch from base 2 to the natural base e. Thus we define

$$H_e(p_0,p_1,\ldots,p_{L-1}) \doteq -\sum_{i=0}^{L-1} p_i \ln p_i = \ln 2 \times H(p_0,p_1,\ldots,p_{L-1}) \qquad (17)$$

Some further interesting properties then are

(4)

$H(p,1-p)$ *is a convex function of* p *and attains its maximum at* $p = \frac{1}{2}$

Proof: Since $H(p,1-p)$ is a linearly scaled version of $H_e(p,1-p)$, we only need to prove this property for the latter. Now

$$H_e(p,1-p) - H_e\left(\frac{1}{2},\frac{1}{2}\right) = p \ln \frac{1/2}{p} + (1-p) \ln \frac{1/2}{1-p}$$
$$\leq p\left(\frac{1/2}{p} - 1\right) + (1-p)\left(\frac{1/2}{1-p} - 1\right)$$
$$= 0$$

The inequality is due to the fact that $\ln x \leq x - 1$. This proves the location of the maximum.

To prove convexity we need to show that for all p_1,p_2, $0 < \theta < 1$, and $p = \theta p_1 + (1 - \theta)p_2$,

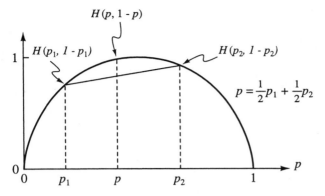

Figure 7.2
Convexity of function $H(p, 1 - p)$

$$\theta H_e(p_1, 1 - p_1) + (1 - \theta)H_e(p_2, 1 - p_2) - H_e(p, 1 - p)$$

is at most 0 (see figure 7.2 for $\theta = \frac{1}{2}$). But the above quantity is equal to (the following inequality is again a result of the fact that $\ln x \le x - 1$)

$$\theta\left[p_1 \ln \frac{p}{p_1} + (1 - p_1)\ln \frac{1 - p}{(1 - p_1)}\right] + (1 - \theta)\left[p_2 \ln \frac{p}{p_2} + (1 - p_2)\ln \frac{1 - p}{(1 - p_2)}\right]$$

$$\le \theta\left[p_1\left(\frac{p}{p_1} - 1\right) + (1 - p_1)\left(\frac{1 - p}{(1 - p_1)} - 1\right)\right]$$

$$+ (1 - \theta)\left[p_2\left(\frac{p}{p_2} - 1\right) + (1 - p_2)\left(\frac{1 - p}{(1 - p_2)} - 1\right)\right]$$

$$= 0$$

Q.E.D.

(5)

$H(p_0, p_1, \ldots, p_{M-1})$ *is a convex function of the probability distribution*[7] $\mathbf{p} = (p_0, p_1, \ldots, p_{M-1})$ *and attains its maximum if* $p_i = 1/M$ *for all* $i = 0, 1, \ldots, M - 1$

Proof: The proof is omitted; it follows along the same lines as the proof of property **(4)**.

7. In this book, boldface variable always refers to a vector.

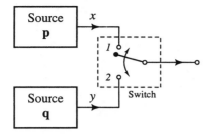

$$\theta = P \{ \text{ switch touches contact 1 } \}$$

Figure 7.3
Source resulting from random sampling of outputs of two sources

- **Interpretation of property (5):**

(A) Uncertainty is maximal when no outcome is more probable than any other.
(B) Letting $H(\mathbf{p}) = H(p_0, p_1, \ldots, p_{M-1})$, we have by definition of convexity that for all $0 < \theta < 1$

$$H(\theta \mathbf{p} + (1 - \theta)\mathbf{q}) \geq \theta H(\mathbf{p}) + (1 - \theta)H(\mathbf{q})$$

with equality if and only if $\mathbf{p} = \mathbf{q}$

The above inequality can be interpreted as meaning that if we have two sources that are sampled by a statistical switch (see figure 7.3), then uncertainly increases if the switch position is made invisible to the observer.

(6)

$H(p_0, p_1, \ldots, p_{M-1}) \geq 0$ *with equality if and only if* $p_i = 1$ *for some i.*

Proof:

$$H(p_0, p_1, \ldots, p_{M-1}) = \sum_{i=0}^{M-1} p_i \log \frac{1}{p_i} \geq 0$$

because for $0 \leq p_i < 1$ each term in the sum is nonnegative.[8] Equality holds only if each term in the sum equals 0, which is the case only if $p_i = 1$ for some i and $p_j = 0$ for $j \neq i$.
Q.E.D.

8. In information theory, we conventionally define
$0 \log 0 \doteq \lim_{x \downarrow 0}[x \log x] = 0$

- **Interpretation of property (6):**

Information and/or uncertainty cannot be a negative quantity.

(7)

$$H(p_0, p_1, \ldots, p_{M-1}) \leq -\sum_{i=0}^{M-1} p_i \log q_i$$

with equality if and only if $p_i = q_i$ *for all i.*

Proof:

$$\sum_{i=0}^{M-1} p_i \ln \frac{q_i}{p_i} \leq \sum_{i=0}^{M-1} p_i \left(\frac{q_i}{p_i} - 1 \right) = 0$$

Q.E.D.

- **Interpretation of property (7):**

An observer is more uncertain about the data if he misestimates the distribution governing the data source. (This interpretation will be operationally justified in section 7.6 after we prove the source-coding theorem in section 7.5).

We conclude this section by remarking that the nonnegative quantity

$$D(\mathbf{p}\|\mathbf{q}) \doteq \sum_{i=0}^{M-1} p_i \log \frac{p_i}{q_i} \qquad (18)$$

plays an interesting role in statistics. We will use it in chapter 13 when we discuss the maximum entropy estimation method. $D(\mathbf{p}\|\mathbf{q})$ is referred to by various names, such as *Kullback-Leibler distance* [3], *relative entropy, discrimination, or divergence.*

7.4   An Alternative Point of View and Notation

We can consider the outputs of an information source to be realizations of *random variables*, denoted by X. A random variable is specified by the distribution of its values, and in this case

$$P\{X = i\} = p_i \qquad (19)$$

For a random variable X to be defined, both its realization alphabet $\{0, 1, \ldots, M - 1\}$ and its distribution $\mathbf{p} = (p_0, p_1, \ldots, p_{M-1})$ must be specified. Thus we can denote by $H(X)$ the entropy governed by the distribution $\mathbf{p}$, because the argument X of the function $H(\ )$ implies $\mathbf{p}$ and vice versa.

Consider next a *new* random variable Z that is a function of the output random variable X:

$$Z(X) = -\log P\{X\} \tag{20}$$

Then[9]

$$\mathbf{E}[Z] = \mathbf{E}[-\log P\{X\}] = -\sum_{i=0}^{M-1} p_i \log p_i = H(X)$$

So we see that entropy is an average, and we can interpret[10] $-\log p_x$ as the amount of information provided by the particular output x.

The entropy corresponding to output sequences $x_1, x_2, \ldots, x_k$ is then

$$H(X_1, X_2, \ldots, X_k) = -\sum_{x_1,\ldots,x_k} P\{X_1 = x_1, X_2 = x_2, \ldots, X_k = x_k\}$$

$$\times \log P\{X_1 = x_1, X_2 = x_2, \ldots, X_k = x_k\}$$

and if the source is constant and memoryless, so that it satisfies

$$P\{X_1 = x_1, X_2 = x_2, \ldots, X_k = x_k\} = \prod_{i=1}^{k} P\{X_i = x_i\} = \prod_{i=1}^{k} p_{x_i}$$

then

$$H(X_1, X_2, \ldots, X_k) = kH(X) \tag{21}$$

In general, we are interested in per-output entropy defined by

$$H = \lim_{k \to \infty} \frac{1}{k} H(X_1, X_2, \ldots, X_k) \tag{22}$$

With the new notation, we have

$$H(X_1, X_2, \ldots, X_k) - \sum_{i=1}^{k} H(X_i)$$

$$= \sum_{x_1,\ldots,x_k} P\{X_1 = x_1, \ldots, X_k = x_k\} \log \frac{P\{X_1 = x_1\} \ldots P\{X_k = x_k\}}{P\{X_1 = x_1, \ldots, X_k = x_k\}} \tag{23}$$

$$\leq 0$$

9. $\mathbf{E}[\ ]$ denotes the expectation operation.

10. This is just an intuitive interpretation. Unlike entropy itself, it has no operational justification.

So we see that

Entropy is maximal when the random variables are independent.

We can further define *conditional* entropy

$$H(X|Y) \doteq - \sum_{x,y} P\{X = x, Y = y\} \log P\{X = x | Y = y\}$$

$$= \sum_{y} P\{Y = y\} H(X|Y = y) \tag{24}$$

and conclude that

Conditional entropy is an average of entropies.

Finally,

$$H(X, Y) = - \sum_{x,y} P\{X = x, Y = y\} \log P\{X = x, Y = y\}$$

$$= - \sum_{x,y} P\{X = x, Y = y\} [\log P\{X = x\}$$

$$+ \log P\{Y = y | X = x\}] \tag{25}$$

$$= H(X) + H(Y|X)$$

Relation (25) has the following interpretation:

Uncertainty about two random variables equals the uncertainty about the first variable plus the average uncertainty about the second variable when the first variable is known.

Relations (24) and (25) can be straightforwardly generalized to multiple random variables, that is:

$$H(X_1, \ldots, X_k | Y_1, \ldots, Y_n)$$

$$= \sum_{y_1, \ldots, y_n} P\{Y_1 = y_1, \ldots, Y_n = y_n\} H(X_1, \ldots, X_k | Y_1 = y_1, \ldots, Y_n = y_n) \tag{26}$$

and

$$H(X_1, \ldots, X_k, Y_1, \ldots, Y_n)$$
$$= H(X_1, \ldots, X_k) + H(Y_1, \ldots, Y_n | X_1, \ldots, X_k) \tag{27}$$

A property that can be demonstrated by the reasoning that led to (25) is

$$H(X_1, \ldots X_l, X_{l+1}, \ldots, X_k | Y_1, \ldots, Y_n)$$

$$= H(X_1, \ldots, X_l | Y_1, \ldots, Y_n) + H(X_{l+1}, \ldots, X_k | Y_1, \ldots, Y_n, X_1, \ldots, X_l)$$

$$(28)$$

Its repeated application gives

$$H(X_1, \ldots, X_k) = H(X_1) + \sum_{i=2}^{k} H(X_i | X_1, \ldots, X_{i-1}) \tag{29}$$

Finally, the inequality $\ln x \le x - 1$ results in

$$H(X_1 | X_2, \ldots, X_k, X_{k+1}) \le H(X_1 | X_2, \ldots, X_k)$$

with equality if and only if X_1 and X_{k+1} are independent when conditioned on $X_2, \ldots, X_k$

$$(30)$$

Property (30) can be pleasingly interpreted by:

Uncertainty decreases when more information is known.

7.5 A Source-Coding Theorem

In section 7.4, we derived the mathematical formula for an information measure, entropy, using as a basis our intuition to tell us what properties such a measure should possess. Now we would like to confirm the correctness of the measure by showing that it arises naturally from yet another point of view: encoding of information generated by data sources.

Consider again the information source of figure 7.1 with output alphabet $\mathscr{X} = \{0, 1, \ldots, M - 1\}$. We would like to *encode* its outputs into strings of binary symbols.[11] The code words we will use will in general be of varying length, and we want to answer the question:

What is the minimum number L of bits that must be used on the average to encode an output symbol of the information source?

The idea is that this minimum number L measures the source's information content. The smaller L is, the less information the source contains. Furthermore, L measures the information content in concrete

---

11. Everything we prove here can be generalized to encoding into an N-ary alphabet. We omit dealing in that generality because our purpose is strengthening the interpretation of the entropy function, not the development of source codes.

Source
output
symbol   Codeword

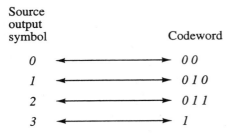

Figure 7.4
Variable length encoding of source outputs of a quaternary source

units: bits.[12] We will see that L will be related to the entropy of the source, H.

Figure 7.4 is a concrete example of what we have in mind. It shows a possible encoding of source outputs from an alphabet of size $M = 4$.

Now a variable length code must be *uniquely decipherable*; that is, for any given string of code bits produced, it must be possible to decide which source symbols "caused it." No commas between code words are allowed: Commas are also a symbol, hence we would be encoding in a ternary alphabet if we used them. The code of figure 7.4 is uniquely decipherable. As an example, the code sequence 010011101000 corresponds to the unique source sequence 12310. In fact, the only way the initial subsequence can be generated is as a result of source output 1. Then the only way the next subsequence 011 can be generated is as a code word of the source symbol 2. The next digit, 1, results uniquely from source symbol 3, etc.

For a code to be uniquely decipherable, it is sufficient that it satisfy the *prefix condition*:

No code word is a prefix to another code word[13]

A code satisfies the prefix condition if it is a *tree code*. The code of figure 7.4 can be diagrammed in the form of the tree in figure 7.5. Here,

12. That is, binary symbols.

13. This is a sufficient but not a necessary condition. A code satisfying this condition is called *instantaneous*, because as soon as the decoder observes that a sequence equal to a code word has been generated, it can conclude that it is that code word, since the continuation of this sequence must result from a start of another code word.

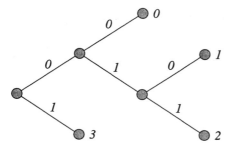

Figure 7.5
Tree representation of the code of figure 7.4

taking the top branch of the tree out of a node results in code symbol 0, and taking the bottom branch results in a 1.

The tree representation of a code makes decoding conceptually very easy. A path from the tree root to any given leaf of the tree represents a code sequence; that particular sequence encodes the symbol attached to that leaf of the tree. At the start of the process, the decoder places itself at the root. It traces the path as directed by the individual code symbols. Once a leaf is reached, the source output symbol associated with the leaf is decoded and the decoder places itself again at the root of the tree, and so forth.

To find the minimum average code length L, we need to introduce some terminology. Let x denote a source output, let $P(x)$ be its probability, and let $l(x)$ be the length of the binary code word assigned to x. Then

$$L \doteq \sum_x l(x)P(x) \tag{31}$$

is the average code length.

LEMMA 7.1 THE KRAFT-SZILARD INEQUALITY [4]. For $l_1, l_2, \ldots, l_k$ to be the lengths of paths (in branches) from the root to different leaves of a tree, it is necessary and sufficient that

$$\sum_{i=1}^{k} 2^{-l_i} \leq 1 \tag{32}$$

Proof: Choose a length $N \geq l_i$ for all i, and construct a full tree of depth N (see figure 7.6, where $N = 4$). Then from a node at depth l_i from the root, there grow exactly 2^{N-l_i} leaves of the full tree. So the total number

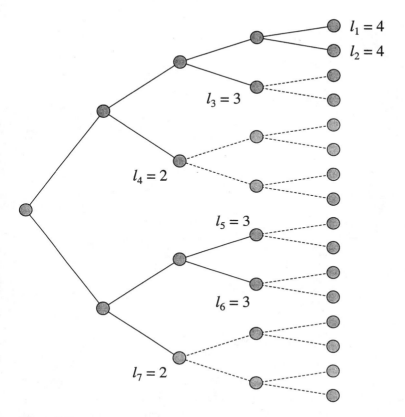

Figure 7.6
Full tree of depth $N = 4$ and the placement of code words of lengths $\{4, 4, 3, 2, 3, 3, 2\}$ into it

of leaves growing out from nodes at depths $l_1, l_2, \ldots, l_k$ is equal to

$$\sum_{i=1}^{k} 2^{N-l_i}$$

But if these nodes lie on different paths from the root, then the sum above cannot exceed the total number 2^N of leaves of a full tree of depth N. Thus if the code lengths $l_1, l_2, \ldots, l_k$ can be implemented in a tree code, inequality

$$\sum_{i=1}^{k} 2^{N-l_i} \le 2^N$$

and therefore (32) must hold. This proves necessity.

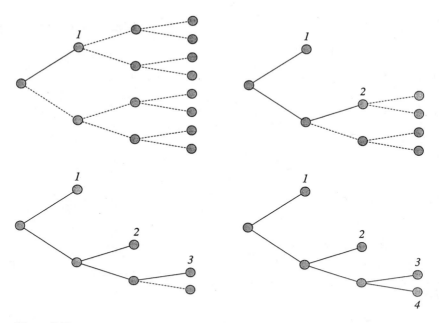

Figure 7.7
Construction of a code of word lengths $\{1, 2, 3, 3\}$

The inequality (32) is also sufficient, because we can create the code as follows: Arrange code words in increasing length, that is, $l_1 \le l_2 \le \ldots \le l_k$. Occupy the topmost node at depth l_1 with the first code word, and erase from the full tree the growth leaving that node. (We erase a total of 2^{N-l_1} leaves as well as all the branches leading to these leaves from the node at depth l_1.) Occupy the topmost available node at depth l_2 with the second code word, and erase from the full tree the growth leaving that node. We assert that if (32) holds then it is possible to continue in this manner until an available node at depth l_k has been found and occupied by the last code word (Figure 7.7 illustrates this construction process for code lengths 1,2,3,3). For assume the contrary and suppose that for some $m \le k$ no node at depth l_m remains available. That means that the number of nodes erased or occupied at depth l_m in the code construction process exceeded or equaled the total number of nodes originally existing at depth l_m (that is, 2^{l_m}) before this step was tried. The number of nodes at depth l_m erased up to this point is exactly

$$\sum_{i=1}^{m-1} 2^{l_m - l_i}$$

so we must have

$$\sum_{i=1}^{m-1} 2^{l_m - l_i} \geq 2^{l_m}$$

which contradicts condition (32).
Q.E.D.

We will now examine just how short the average length L can be.

Let $\lceil y \rceil$ denote the least integer greater than or equal to y. Let us make the code length assigned to source output x equal to

$$l(x) = \lceil -\log P(x) \rceil < 1 - \log P(x) \tag{33}$$

Then

$$-l(x) \leq \log P(x)$$

and therefore

$$\sum_x 2^{-l(x)} \leq \sum_x 2^{\log P(x)} = \sum_x P(x) = 1$$

so that the code length assignment satisfies the Kraft-Szilard inequality and the code is thus implementable.

Now by the choice (33), the average length satisfies

$$L = \sum_x l(x) P(x) < \sum_x (1 - \log P(x)) P(x) = 1 + H(X) \tag{34}$$

and L thus exceeds the value of entropy by less than 1.

L can be made as close to entropy as desired by *block coding*: Instead of encoding single output symbols of the source, we will encode their blocks of length n, that is sequences $x_1, x_2, \ldots, x_n$. Since the source is assumed to be memoryless, it follows directly from (34) and (21) that the average length of this block code satisfies

$$L < 1 + H(X_1, X_2, \ldots, X_n) = 1 + nH(X)$$

so that codes exist that provide an average number of bits per source output, $\frac{1}{n} L$, satisfying

$$\lim_{n \to \infty} \frac{1}{n} L \leq H(X) \tag{35}$$

The converse to (35) can also be proven [5], that is, that

$$No \; code \; exists \; for \; which \; \frac{1}{n} L < H(X) \; for \; any \; n. \tag{36}$$

Inequalities (35) and (36) together constitute *Shannon's First Coding Theorem* [1]. It allows us to give a clear operational interpretation to the concept of entropy:

THEOREM 7.1   SHANNON'S FIRST CODING THEOREM.   The entropy of an information source is equal to the minimum average number of bits per symbol that must (and can in the limit) be used to encode source outputs: It is the information content of the source measured in bits.

7.6   A Brief Digression

We are now in a position to justify the interpretation of entropy property (7) stated in section 7.3,

$$H(p_0, p_1, \ldots, p_{M-1}) \le - \sum_{i=0}^{M-1} p_i \log q_i$$

If the encoder "thinks" that the source being encoded has distribution $\mathbf{q} = \{q_0, \ldots, q_{M-1}\}$, then it will assign to the ith symbol a code word of length

$$l_i = \lceil -\log q_i \rceil$$

and as a result, its code will have an average length that will exceed $- \sum_{i=0}^{M-1} p_i \log q_i$, whereas had the encoder known the true distribution $\mathbf{p}$, it could have constructed a code with an average code word length arbitrarily close to $H(p_0, p_1, \ldots, p_{M-1})$. Thus misestimation of the source distribution is costly.

7.7   Mutual Information

We can now introduce another important concept of information theory: *mutual information*.

Consider figure 7.8. It contains a memoryless source governed by the distribution $P(x)$ observed through a *noisy channel*[14] characterized by the

---

14. The channel is called *noisy* because, in general, it does not allow the receiver observing the outputs y to determine uniquely the corresponding inputs x.

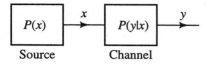

Source   Channel

Figure 7.8
Direct transmission of source symbols through a noisy channel

distributions $P(y|x)$. How much information about the source output x does an observer gain by knowing the channel output y?

One can reason as follows: Before the observer sees y, his uncertainty about x is $H(X)$. After observing the channel output, his uncertainty is reduced, on the average, to $H(X|Y)$. An observer therefore gains an average amount of information

$$I(X; Y) \doteq H(X) - H(X|Y) \tag{37}$$

for each channel output symbol observed.

The function $I(X; Y)$ is called the *average mutual information* between X and Y. We use the term *between* because $I(X; Y)$ is a symmetrical function of these random variables. In fact, we get from (37) that[15]

$$I(X; Y) \doteq H(X) - H(X|Y) = \sum_{x,y} P(x,y) \log \frac{P(x|y)}{P(x)}$$

$$= \sum_{x,y} P(x,y) \log \frac{P(x|y)P(y)}{P(x)P(y)} = \sum_{x,y} P(x,y) \log \frac{P(x,y)}{P(x)P(y)} \tag{38}$$

$$= H(X) + H(Y) - H(X, Y) = \sum_{x,y} P(x,y) \log \frac{P(y|x)}{P(y)}$$

$$= H(Y) - H(Y|X)$$

It is a direct result of (30) that $0 \le H(X|Y) \le H(X)$, and so

$$0 \le I(X, Y) \le \min\{H(X), H(Y)\} \tag{39}$$

with equality on the left-hand side if and only if X and Y are independent, and equality on the right-hand side if and only if either X determines Y or Y determines X.

15. It is interesting to note from (38) that the average information function is related to the divergence defined in (18). In fact,
$I(X; Y) = D(\mathbf{P}\|\mathbf{Q})$
where $\mathbf{P} \doteq P(x,y)$ and $\mathbf{Q} \doteq P(x)P(y)$.

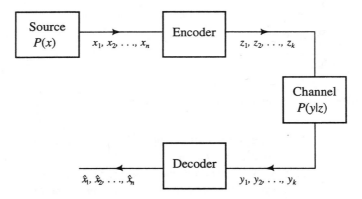

Figure 7.9
Transmission of encoded information through a noisy channel

Note from (38) that

$$I(X; Y) = \mathbf{E}\left[\log \frac{P\{X, Y\}}{P\{X\}P\{Y\}}\right] \tag{40}$$

and so it is possible to interpret[16] the quantity $\log P(x, y)/P(x)P(y)$ as *the mutual information between symbols x and y*.

The (average) mutual information function can also be supported operationally by a coding theorem. We will omit its proof but will state it below.

Define the *capacity* of a given (noisy) channel (specified by the distribution $P(y|x)$) by

$$C \doteq \max_{P(x)} I(X; Y) \tag{41}$$

So C is obtained by varying the output distribution of the information source (i.e., the input distribution of the channel) until one is found that maximizes the mutual information function for the given channel.

Figure 7.9 diagrams a block-coding scheme for transmission through the channel. The *encoder* has a buffer that accumulates sequences $\mathbf{x} = x_1$, $x_2, \ldots, x_n$ of source outputs and maps them into sequences $z_1, z_2, \ldots, z_k$ of

16. As we have done before, we caution the reader that this is only an intuitive interpretation. Unlike average mutual information $I(X; Y)$, which is justified by the coding theorem below, the concept of mutual information between x and y has no such operational justification.

channel inputs. At the output of the channel, the *decoder* accumulates sequences $y_1, y_2, \ldots, y_k$ and maps them into sequences $\hat{\mathbf{x}} = \hat{x}_1, \hat{x}_2, \ldots, \hat{x}_n$ of estimates of the source outputs. The aim is to have the probability $P\{\hat{\mathbf{x}} = \mathbf{x}\}$ of correct decoding close to 1 while the ratio $R \doteq \frac{n}{k}$ of the two block lengths (called the *coding rate*) is as large as possible (the larger R is, the fewer symbols are sent over the channel per source output, resulting in a parsimonious use of the channel).

We then have

THEOREM 7.2   SHANNON'S SECOND CODING THEOREM [1].   If the normalized (by the rate R) source entropy satisfies

$$R \times H(X) < C$$

then for any $\varepsilon > 0$ it is possible to find an encoder-decoder pair of rate R (using a sufficiently large block length n) such that

$$P(\hat{\mathbf{X}} \neq \mathbf{X}) < \varepsilon$$

If $R \times H(X) > C$ then for all codes there exists a $\delta > 0$ such that

$$P(\hat{\mathbf{X}} \neq \mathbf{X}) > \delta$$

7.8   Additional Reading

The text [2] is the current favorite reference for learning all aspects of basic information theory. The simplest book to read, if still available, is that by Norman Abramson [6].

References

[1] C.E. Shannon, "A mathematical theory of communication," *Bell System Technical Journal*, vol. 27, pp. 379–423 and 623–56, 1948.

[2] T.M. Cover and J.A. Thomas, *Elements of Information Theory*, John Wiley & Sons, New York, 1991.

[3] S. Kullback and R.A. Leibler, "On information and sufficiency," *Ann. Math. Stat.*, vol. 22: pp. 79–86, 1951.

[4] L.G. Kraft, *A device for quantizing, grouping, and coding amplitude modulated pulses*, Master's Thesis, Department of Electrical Engineering, Massachusetts Institute of Technology, Cambridge, 1949.

[5] R.M. Fano, *Class Notes for Transmission of Information*, Course 6.574, Massachusetts Institute of Technology, Cambridge, 1952.

[6] N. Abramson, *Information Theory and Coding*, McGraw-Hill, New York, 1963.

Chapter 8

The Complexity of Tasks—
The Quality of Language
Models

8.1 The Problem with Estimation of Recognition Task Complexity

It is obviously of interest to be able to compare the complexities (i.e., difficulties) of different recognition tasks. Information theory provides a tool: entropy. The problem is how to apply this measure to speech recognition.

From the source-channel formulation of the recognition problem (see figure 8.1) we might at first conclude that the task's complexity is given by the entropy of the source, $H(\mathbf{W})$. This does not, however, take into account any acoustic aspects, such as the speaker's variability, the vocabulary's acoustic similarity or the medium's noise and transmission characteristics. The source-channel formulation is useful in that it partitions the recognizer tasks into components[1] and defines the corresponding statistical variables, but the difficulty of recognition cannot be unequivocally estimated from the statistics of what our formulation calls the source.

The input to the recognizer is really the incoming signal, denoted by $\mathbf{B}$, which is influenced by the characteristics of the word source, the speaker mechanism, and the transmission medium. But the complexity of the task is not measured by $H(\mathbf{B})$ either, because the recognition task is not the restoration of $\mathbf{B}$ but the estimation of $\mathbf{W}$ given the observed signal $\mathbf{B}$. So the straightforward approach according to information theory would be to estimate the *average conditional entropy* $H(\mathbf{W}|\mathbf{B})$, which measures the uncertainty about the source output $\mathbf{W}$ when the acoustics $\mathbf{B}$ have been observed.

---

1. See section 1.3.

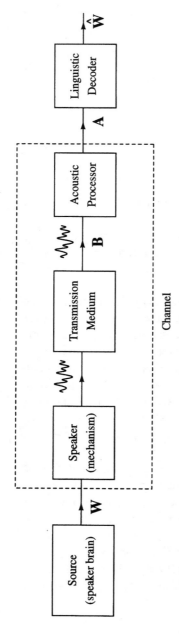

Figure 8.1
A schematic diagram of the speech generation and recognition process

Some thought will reveal that we cannot realize such an ambition, for knowing $H(\mathbf{W}|\mathbf{B})$ implies knowing the statistics $P(\mathbf{W},\mathbf{B})$. What would be the consequence of being in the possession of these statistics? Since the recognition task is to find the maximizing argument $\hat{\mathbf{W}}$ of the product $P(\mathbf{A}|\mathbf{W})P(\mathbf{W})$, and since

$$P(\mathbf{A}|\mathbf{W}) = \sum_{\mathbf{B}} P(\mathbf{A}|\mathbf{B})P(\mathbf{B}|\mathbf{W})$$

then the recognition problem would consist of finding the probability $P(\mathbf{A}|\mathbf{B})$ corresponding to the performance of the designer's acoustic processor and of carrying out an efficient hypothesis search for $\hat{\mathbf{W}}$.

In reality, all we have at our disposal are estimates of the probabilities $P(\mathbf{A}|\mathbf{W})$ and $P(\mathbf{W})$ implicit in the models we use (e.g., HMMs for $P(\mathbf{A}|\mathbf{W})$ and trigrams for $P(\mathbf{W})$), and we are certainly not prepared to say that he has a harder task who has a higher conditional entropy $H(\mathbf{W}|\mathbf{A})$, that is, he who perhaps works on the basis of worse models! In fact, a recognizer could conceivably use a really bad model $P(\mathbf{A}|\mathbf{W})$ that would result in $H(\mathbf{W}|\mathbf{A}) \cong H(\mathbf{W})$.

Since $H(\mathbf{W})$ is the only conceivable upper bound on both $H(\mathbf{W}|\mathbf{A})$ and $H(\mathbf{W}|\mathbf{B})$, the speech recognition community has been, after all, forced to fall back on estimating $H(\mathbf{W})$ and on using it as a rule-of-thumb measure of task difficulty.

It is then clear that by relying on $H(\mathbf{W})$ we will, at best, be able to compare only those recognition tasks whose acoustics are assumed identical.[2] We will also be able to use information theoretic heuristics to say which of two systems working on the same task has models providing it with a lower $H(\mathbf{W}|\mathbf{A})$ and may thus be expected to have a lower error rate.

8.2 The Shannon Game

In one of his earliest papers [1] Shannon devised an ingenious way to estimate the entropy of English (or of any other language, for that matter). He realized the impossibility of directly and reliably estimating the required statistics and reasoned that they are best embedded in the minds of native speakers. So the question was how to tease these statistics out of those minds. His answer is in the form of a guessing game.

2. That is, the transmission medium, the phonetic similarity of words in the vocabulary, the mix of intended speakers, and so forth.

The basic idea is to have a human subject guess sequentially the words of a text hidden from him and to use the relative frequencies of his guesses as estimates of the probability distribution underlying the source of the text. But the guesses cannot be carried out a word at a time, since there are too many words to choose from (the English vocabulary is too large). So Shannon let his subjects guess letters l_i (including punctuation and space) while revealing to them the entire past. But to estimate $P(l_i|l_1, \ldots, l_{i-1})$ from the relative frequency $f(l_i|l_1, \ldots, l_{i-1})$ of a subject's guesses is still impossible. Just to accumulate enough data for even a moderately large value of i would require an effort that is not practicable.

This is how Shannon got around the problem: Suppose the subject is shown the past $l_1, \ldots, l_{i-1}$ and is allowed to keep guessing the letter l_i until he succeeds. Let r_i be the order of the successful guess (i.e., it took the subject r_i guesses to specify correctly the actual letter l_i). If we regard the human subject as a *constant machine*, then we can expect that when conditioned on $l_1, \ldots, l_{i-1}$, he will always guess l_i on the r_i^{th} try. So the string $r_1, \ldots, r_i, \ldots$ is a perfect encoding of the string $l_1, \ldots, l_i, \ldots$ and vice versa. Consequently the following two entropies have the same value:

$$H(l_1, \ldots, l_i) = H(r_1, \ldots, r_i)$$

Furthermore, with K the size of the letter alphabet (which includes space and punctuation),[3]

$$H(r_1, \ldots, r_i) \leq iH(r) = -i \sum_{r=1}^{K} P(r) \log P(r) \tag{1}$$

and so we can get an upper bound on the desired quantity by estimating $P(r)$. And we can surely process enough data so as to make the relative frequency $f(r)$ a reliable estimate of $P(r)$.

In principle, we can make an analogue of the bound (1) tighter by estimating the relative frequencies of order bigrams $f(r_1, r_2)$ etc., provided the subject has enough patience (and we the money to pay him for it) to generate the required amount of data. Shannon also derived a lower bound on $H(r_1, \ldots, r_i)$ that allowed him to estimate the entropy of English at approximately 1 bit per letter.

We have presented the Shannon approach because of its ingenuity, not because we advocate its use in the measurement of recognition task complexities. The method has at least three problems:

3. See property **(7)** in section 7.3.

1. Is the text that the subjects guessed really representative of the recognition task? (Certainly Shannon's was not representative of all English)
2. Is there enough text so that it provides examples of all the variations of which the task is capable? (Again, Shannon did not have enough.)
3. Are the human subjects sufficiently familiar with the recognition domain, and are they putting enough effort into their guessing, so that the order statistics can be taken seriously?

Through the years, several improvements have been made to the Shannon scheme [2], none of which are really practical enough to be used for our purposes. A very interesting and radically different string coding approach due to Ziv and Lempel [3] does exist that can be proven to converge to the entropy of the process. But unfortunately, its convergence rate has turned out to be too slow to make it practicable.

8.3 Perplexity

In the preceding chapter on information theory we have proven the inequality (see property (6) in section 7.3)

$$H(\mathbf{p}) \le - \sum_i p_i \log q_i \tag{2}$$

and in section 7.6 we interpreted the right-hand side of (2) as the entropy from the point of view of the user who has misestimated the source distribution to be $\mathbf{q}$ rather than $\mathbf{p}$, which it really is. So an analogue of $-\sum_i p_i \log q_i$ would represent the task difficulty from the recognizer's point of view. This analogue is called the *logprob* (*LP*) and is defined by

$$LP \doteq \lim_{n \to \infty} -\frac{1}{n} \sum_{i=1}^{n} \log Q(w_i | w_1, \ldots, w_{i-1}) \tag{3}$$

where $Q(w_i | w_1, \ldots, w_{i-1})$ denotes the recognizer's estimate of the text production probabilities that is embedded in the recognizer's language model.

LP must be estimated over a test text $w_1, \ldots, w_n, \ldots$ that is separate from that used to estimate the probabilities $Q(w_i | w_1, \ldots, w_{i-1})$. The well-known *law of large numbers* assures that (3) is truly an analogue of (2), because (3) is the *time average* of the *appearance* of the quantity $\log Q(w_i | w_1, \ldots, w_{i-1})$ in text, and the right-hand side of (2) is its *expectation*.

In speech recognition's early days researchers talked about the recognition task's *branching factor*, an estimate of the size of the word list from which a recognizer was choosing when it was deciding which word was spoken. We can turn to information theory to provide us with the definition of *perplexity* that would be the correct analogue of the branching factor:

$$PP \doteq 2^{LP} \tag{4}$$

In fact, the value of perplexity is equal to the size of an imaginary *equivalent list*, one whose words are equally probable. For such a list of size K, $LP = \log K$, and so for that case we get the consistent result $PP = K$.

Let us repeat that although perplexity measures the text source's complexity from the recognizer's point of view, it does not measure the task difficulty, because it takes no account of acoustics and because it is based on the language model's estimate of the time statistics of the source.

8.4 The Conditional Entropy of the System

To estimate the recognition system's overall quality, we could try to estimate the value of the conditional entropy function by its time average.[4] The appropriate formula (applied to data that was not used to estimate the values of $P(\mathbf{A}|\mathbf{W})$ or $P(\mathbf{W})$) then is

$$H_{\mathbf{A}} \doteq \lim_{n\to\infty} -\frac{1}{n} \log \frac{P(\mathbf{A}|\mathbf{W})P(\mathbf{W})}{P(\mathbf{A})} \tag{5}$$

where n is the number of words in the string $\mathbf{W}$.

Presumably expression (5) would be evaluated by letting a subject read the text $\mathbf{W}$ into an acoustic processor that would produce the output string $\mathbf{A}$. The probability $P(\mathbf{A}|\mathbf{W})$ would be computed by creating the HMM corresponding to the text $\mathbf{W}$ and carrying out the standard forward algorithm.[5] The language model would provide the value of $P(\mathbf{W})$. This leaves the problem of how to estimate $P(\mathbf{A})$. A brute-force method based on the formula

$$P(\mathbf{A}) = \sum_{\mathbf{W'}} P(\mathbf{A}|\mathbf{W'})P(\mathbf{W'})$$

is, of course, out of the question, since the number of elements in the sum exceeds all reasonable bounds.

---

4. Similarly to (3), which gives us the time average estimate of entropy.

5. Recursions (9) and (20) in sections 2.3 and 2.6.

Since H_A should correspond to the system's average per-word uncertainty about what was spoken when the acoustics were observed, a possible approximation can be obtained by creating an HMM whose basic building blocks would be the HMMs corresponding to the vocabulary's various words. The latter would be interconnected by null transitions whose probabilities would be those the language model provided. $P(\mathbf{A})$ would then again be evaluated by running the forward algorithm based on the HMM created.

The situation is shown schematically in figure 8.2, where the states correspond to the L possible language model equivalence classes $\Phi_1, \Phi_2, \ldots, \Phi_L$ whose identity is determined by the *mapping* $\Phi(\mathbf{h}_i) = \phi(w_1, \ldots, w_{i-1})$ performed by the switch on the diagram's left. The transition from state $\Phi(\mathbf{h}_i)$ to state $\Phi(\mathbf{h}_i, w_i)$ leads through the HMM of the

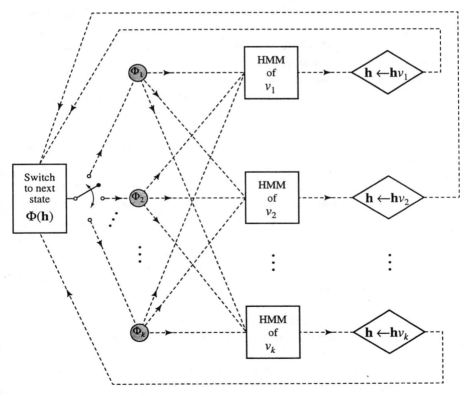

Figure 8.2
A composite HMM trained to model the generation of the symbol sequences A

word w_i (the vocabulary is of size $k = |\mathcal{V}|$) and is taken with probability $P(w_i|\Phi(\mathbf{h}_i))$.

It is obvious that, under this scheme, evaluating H_A would be just as complicated as running an entire recognizer with a Viterbi hypothesis search component. But knowing the value of H_A is of practical interest only if it predicts the recognizer performance, or improvements in it, so one is willing to pay only a limited price for finding H_A. Thus if its computation is to have any value, the calculation of $P(\mathbf{A})$ must be simplified by accepting an approximation.

What could be done is to train some relatively simple production model of output strings $\mathbf{A}$. One possible way might be to create a fully connected K state HMM and train it on some sample data $\mathbf{A}$. A more plausible approach would be to construct the model on a phonetic basis:

1. Train phonetic HMMs, one for each member of the phonetic alphabet.
2. Replacing words in text by phonetic base forms, obtain a long phone string. Use it to create a language model of phone trigrams, $P(\varphi_3|\varphi_1, \varphi_2)$.
3. Create an HMM whose basic building blocks are the HMMs corresponding to the phonetic alphabet's various elements, interconnected by null transitions whose probabilities are set to $P(\varphi_3|\varphi_1, \varphi_2)$.
4. Train the constructed HMM on sample acoustic processor output data, reestimating all its statistical parameters using the previous step's parameters as initial values.
5. Use the forward algorithm[6] to compute $P(\mathbf{A})$ based on the trained HMM.

Although the approximate H_A computed in this manner will not necessarily estimate with accuracy the conditional entropy associated with the system, it will facilitate a valid comparison between different systems, provided the structure of the HMMs used as the basis of the $P(\mathbf{A})$ computation is the same for all systems being compared.

8.5 Additional Reading

The Cover and King article [2] estimates text entropy somewhat differently than does Shannon's method described in section 8.2. The players, instead of guessing in sequence at letters, bet fractions of their accumulated capital. The Nemetz method [4] is entirely distinct and provides an upper bound on the entropy.

---

6. Recursions (9) and (20) in sections 2.3 and 2.6.

Mercer and colleagues [5] tried to create the best (essentially) trigram language model for unlimited vocabulary text, measured its perplexity, and challenged the world to do better.

Feretti, Maltese, and Scarci define a new concept, *speech decoder entropy*, which attempts to measure the quantity H_A defined in (5). They describe their method and present their result in reference [6].

References

[1] C.E. Shannon, "Prediction and entropy of printed English," *Bell System Technical Journal*, vol. 30, pp. 50–64, 1951.

[2] T.M. Cover and R. King, "A convergent gambling estimate of English," *IEEE Transactions on Information Theory*, vol. IT-24, pp. 413–21, 1978.

[3] J. Ziv and A. Lempel, "A universal algorithm for sequential data compression," *IEEE Transactions on Information Theory*, vol. IT-23, pp. 337–43, 1997.

[4] T. Nemetz, "Entropy estimation via reconstruction of mutilated texts," *Colloques Internationaux du CNRS: Theorie de l'Information*, no. 276, 4–8 juillet, 1977.

[5] P.F. Brown, S.A. Della Pietra, V.J. Della Pietra, J.C. Lai, R.L. Mercer, "An estimate of an upper bound for the entropy of English," *Computational Linguistics*, vol 18, no. 1, pp. 31–40, March 1992.

[6] M. Ferretti, G. Maltese, S. Scarci, "Measuring information provided by language model and acoustic model in probabilisitic speech recognition: theory and experimental results," *Speech Communication*, vol. 9, nos. 5/6, pp. 531–39, December 1990.

Chapter 9

The Expectation-
Maximization Algorithm and
Its Consequences

9.1 Introduction

In this chapter we will derive the Expectation-Maximization (EM) algorithm [1] and use it to justify the convergence and optimality of the Baum-Welch algorithm that estimates maximum likelihood HMM parameter values from data.[1] We will then generalize the Baum-Welch algorithm to apply to continuous, normally distributed acoustics. We will finally introduce the *tied mixture* acoustic model used in many current continuous speech recognizers. The EM algorithm's applicability is, of course, not restricted to the training of HMMs: It is much more general.

The development below is based on a special case of Jensen's inequality, which we already proved in section 7.3.[2]

LEMMA 9.1.   If $p(x)$ and $q(x)$ are two discrete probability distributions, then

$$\sum_x p(x) \log p(x) \geq \sum_x p(x) \log q(x)$$

with equality if and only if $p(x) = q(x)$ for all x.

9.2 The EM Theorem

We will now develop the main theorem. Let y denote observable data. Let $P_{\theta'}(y)$ be the probability distribution of y under some model whose parameters are denoted by θ'.[3] Let $P_{\theta}(y)$ be the corresponding distribu-

---

1. See section 2.7.

2. Property (7) in section 7.3.

3. θ' denotes the totality of all the parameters whose values are needed to specify the distribution $P_{\theta'}$.

tion under a different setting θ of the same parameters. We are interested in developing conditions for which y is more likely under θ than it is under θ'.[4]

Let t be another random variable whose value is determined in the same process (t is therefore governed by the value of the parameters θ' or θ) that generates y (for instance, in the HMM setting, y might correspond to the observed output sequence and t to state or to transition sequences). Then, because $P_{\theta'}(t|y)$ is a probability distribution that sums to 1,

$$\log P_{\theta}(y) - \log P_{\theta'}(y) = \sum_t P_{\theta'}(t|y) \log P_{\theta}(y) - \sum_t P_{\theta'}(t|y) \log P_{\theta'}(y)$$

Since we can multiply by 1 without changing anything, the above is equal to

$$\log P_{\theta}(y) - \log P_{\theta'}(y)$$

$$= \sum_t P_{\theta'}(t|y) \log P_{\theta}(y) \frac{P_{\theta}(t,y)}{P_{\theta}(t,y)} - \sum_t P_{\theta'}(t|y) \log P_{\theta'}(y) \frac{P_{\theta'}(t,y)}{P_{\theta'}(t,y)}$$

$$= \sum_t P_{\theta'}(t|y) \log \frac{P_{\theta}(t,y)}{P_{\theta}(t|y)} - \sum_t P_{\theta'}(t|y) \log \frac{P_{\theta'}(t,y)}{P_{\theta'}(t|y)}$$

$$= \sum_t P_{\theta'}(t|y) \log P_{\theta}(t,y) - \sum_t P_{\theta'}(t|y) \log P_{\theta'}(t,y)$$

$$+ \sum_t P_{\theta'}(t|y) \log P_{\theta'}(t|y) - \sum_t P_{\theta'}(t|y) \log P_{\theta}(t|y)$$

$$\geq \sum_t P_{\theta'}(t|y) \log P_{\theta}(t,y) - \sum_t P_{\theta'}(t|y) \log P_{\theta'}(t,y)$$

where the inequality follows from lemma 9.1. Thus if the last quantity in the above equation is positive, so is the first. We have thus proven

THEOREM 9.1.   If

$$\sum_t P_{\theta'}(t|y) \log P_{\theta}(t,y) > \sum_t P_{\theta'}(t|y) \log P_{\theta'}(t,y) \qquad (1)$$

then

$$P_{\theta}(y) > P_{\theta'}(y) \qquad (2)$$

This is the basic EM theorem. It says that if we start with the parameter setting θ' and find a parameter setting θ for which the inequality (1) holds,

---

4. In that case, θ represents an improvement over θ'.

then the observed data y will be more probable under the regime θ than they were under θ'.

To take best advantage of this hill-climbing theorem, we should thus endeavor to find the setting θ that will maximize the left-hand side of (1). Such setting will, of course, satisfy (1) since both of its sides have the same form, which is maximized by that choice of θ. The reason for the name *Expectation-Maximization* is that we take the expectation of the random variable $\log P_\theta(t, y)$ with respect to the old distribution $P_{\theta'}(t|y)$ and then maximize that expectation as a function of the argument θ. It is an algorithm, because the natural way to run it is to choose an initial value of θ', then compute the maximizing θ, then set θ' to θ, compute a new θ, etc. As the process continues, the value of $P_{\theta'}(y)$ will keep climbing toward a limit, since the upper bound $P_{\theta'}(y) \leq 1$ necessarily applies.

The secret of success in applying the EM algorithm is a judicious choice of the auxiliary variable t that will allow finding the maximum of the expectation on the left-hand side of (1). Such a choice is possible for HMMs.

9.3   The Baum-Welch Algorithm

We will now use the EM theorem to derive the Baum-Welch algorithm [2] for a discrete output alphabet $\mathcal{Y}$.[5] To do this with ease, we will use that formulation of HMMs in which various transitions are deterministically related to the observed outputs (as formulated in section 2.2). That is, a function $Y(t)$ assigns outputs y to transitions t.

We will naturally deal with strings of outputs, denoted by $\mathbf{y} = y_1 y_2 \ldots y_n$, and strings of transitions $\mathbf{t} = t_{i_1 j_1}, t_{i_2 j_2} \ldots t_{i_k j_k}$. Here $k \geq n$ because some of the transitions may be null. The expression $t_{i_l j_l}$ denotes the $(j_l)^{th}$ transition out of state i_l. Since $\mathbf{t}$ defines a path, then necessarily $R(t_{i_l j_l}) = i_{l+1}$.[6] (We are following the notation introduced in section 2.2.) The parameter θ of interest consists of the totality of transition probabilities

$$p_{ij} = P(t_{ij})$$

5. See section 2.7.

6. If $P(\mathbf{t}) > 0$, then our notation is such that the transition j_k out of state i_k leads to state i_{k+1}, or equivalently $L(t_{i_k j_k}) = i_k$ and $R(t_{i_k j_k}) = i_{k+1}$.

that define the HMM in question. Thus to vary θ is equivalent to varying the transition probabilities p_{ij}.

We will wish to find the maximum of $\sum_t P_{\theta'}(\mathbf{t}|\mathbf{y}) \log P_\theta(\mathbf{t}, \mathbf{y})$ with respect to θ, which we will obtain by differentiating with respect to the probabilities p_{ij} and equating the result to 0. We thus get[7]

$$
\frac{\partial}{\partial p_{ij}} \left[\sum_t P_{\theta'}(\mathbf{t}|\mathbf{y}) \log P_\theta(\mathbf{t}, \mathbf{y}) - \sum_m \lambda_m \sum_n p_{mn} \right]
$$

$$
= \sum_t P_{\theta'}(\mathbf{t}|\mathbf{y}) \frac{(\partial/\partial p_{ij}) P_\theta(\mathbf{t}, \mathbf{y})}{P_\theta(\mathbf{t}, \mathbf{y})} - \lambda_i
$$

(3)

Since $\mathbf{t}$ determines $\mathbf{y}$, then either $P_\theta(\mathbf{t}, \mathbf{y}) = 0$ (if $\mathbf{y}$ is incompatible with $\mathbf{t}$), or

$$
P_\theta(\mathbf{t}, \mathbf{y}) = P_\theta(\mathbf{t}) = \prod_{l=1}^{K} p_{i_l j_l}
$$

(4)

If $c_{ij}(\mathbf{t})$ denotes the number of times the transition t_{ij} takes place in the string $\mathbf{t}$, then since

$$
\frac{\partial}{\partial p_{ij}} P_\theta(\mathbf{t}, \mathbf{y}) = \frac{\partial}{\partial p_{ij}} \prod_{l=1}^{k} p_{i_l j_l}
$$

we get

$$
\frac{(\partial/\partial p_{ij}) P_\theta(\mathbf{t}, \mathbf{y})}{P_\theta(\mathbf{t}, \mathbf{y})} = \frac{c_{ij}(\mathbf{t})}{p_{ij}}
$$

(5)

Thus equating (3) to 0 we end up with the equation

$$
\sum_t P_{\theta'}(\mathbf{t}|\mathbf{y}) \frac{c_{ij}(\mathbf{t})}{p_{ij}} = \lambda_i
$$

(6)

or,

$$
p_{ij} = \frac{1}{\lambda_i P_{\theta'}(\mathbf{y})} \sum_t P_{\theta'}(\mathbf{t}, \mathbf{y}) c_{ij}(\mathbf{t}) = \frac{1}{K_i} \sum_t P_{\theta'}(\mathbf{t}, \mathbf{y}) c_{ij}(\mathbf{t})
$$

(7)

where K_i plays the role of a normalizing constant that assures that

---

7. As elsewhere, we use the method of undetermined lagrangian multipliers.

$\sum_j p_{ij} = 1$. But, using the Kronecker delta notation[8] and recalling that $\mathbf{t} = t_{i_1 j_1}, t_{i_2 j_2}, \ldots, t_{i_k j_k}$,

$$c_{ij}(\mathbf{t}) = \sum_{l=1}^{k} \delta(t_{ij}, t_{i_l j_l})$$

so that if t_{ij} is not a null transition then

$$\sum_t P_{\theta'}(\mathbf{t}, \mathbf{y}) c_{ij}(\mathbf{t})$$

$$= \sum_{l=1}^{k} \sum_t P_{\theta'}(\mathbf{t}, \mathbf{y}) \delta(t_{ij}, t_{i_l j_l})$$

$$= \sum_{l=1}^{k} P_{\theta'}(y_1, \ldots, y_{l-1}, s_{l-1} = i) p'_{ij} P_{\theta'}(y_{l+1}, \ldots, y_k | s_l = R(t_{ij}))$$

$$= \sum_{l=1}^{k} \alpha_{l-1}^{\theta'}(i) p'_{ij} \beta_l^{\theta'}(R(t_{ij}))$$

(8)

In (8) we have used the definitions

$$\alpha_l(s) \doteq P(y_1, \ldots, y_l, s_l = s)$$
$$\beta_l(s) \doteq P(y_{l+1}, \ldots, y_k | s_l = s)$$

(9)

introduced in section 2.7.

Similarly, if t_{ij} is a null transition then

$$\sum_t P_{\theta'}(\mathbf{t}, \mathbf{y}) c_{ij}(\mathbf{t})$$

$$= \sum_{l=1}^{k} \sum_t P_{\theta'}(\mathbf{t}, \mathbf{y}) \delta(t_{ij}, t_{i_l j_l})$$

$$= \sum_{l=1}^{k} P_{\theta'}(y_1, \ldots, y_{l-1}, s_{l-1} = i) p'_{ij} P_{\theta'}(y_l, \ldots, y_k | s_{l-1} = R(t_{ij}))$$

$$= \sum_{l=1}^{k} \alpha_{l-1}^{\theta'}(i) p'_{ij} \beta_{l-1}^{\theta'}(R(t_{ij}))$$

(10)

8.

$$\delta(a, b) \doteq \begin{cases} 1 & \text{if } a = b \\ 0 & \text{otherwise} \end{cases}$$

The right-hand sides of (8) and (10) are in fact the contributions the
Baum-Welch algorithm makes to the counters that determine the next
value of p_{ij}. We have thus deduced the validity of the Baum-Welch algorithm from the EM theorem. Of course, the values of $\alpha_l(s)$ and $\beta_l(s)$ are
obtained iteratively by formulas (32) and (34) of chapter 2.

9.4   Real Vector Outputs of the Acoustic Processor

We will now generalize HMMs to the case where their output symbols
are normally distributed vectors of real numbers. To avoid unnecessary
complications, we will develop our reestimation formulas for two-
dimensional vectors. The reader will then immediately accept the obvious
generalization to k dimensions.

9.4.1   Development for Two Dimensions

The general setup is this: We have the usual HMM with (possibly multiple) transitions between states. The non-null transitions generate outputs
that are normally distributed two-dimensional real vectors $\mathbf{y}$. The unknown parameters then are (a) the transition probabilities $p(t)$ satisfying[9]

$$\sum_{t : L(t)=s} p(t) = 1 \qquad \text{for all states } s \tag{11}$$

and (b) the parameters $\mathbf{m}(t)$, and $\mathbf{U}(t)$ of the normal density

$$\mathcal{N}_t(\mathbf{y}) = \frac{1}{2\pi\sqrt{|\mathbf{U}(t)|}} \exp\left\{ -\frac{1}{2}(\mathbf{y} - \mathbf{m}(t))\mathbf{U}(t)^{-1}(\mathbf{y} - \mathbf{m}(t))' \right\} \tag{12}$$

where $(\mathbf{y} - \mathbf{m}(t))'$ denotes the transpose of $(\mathbf{y} - \mathbf{m}(t))$. $\mathbf{U}(t)$ is the process's
covariance matrix. To simplify the notation, we will frequently omit the
argument t and use it only when necessary to avoid confusion.

   If

$$\mathbf{U} = \begin{bmatrix} (\sigma_1)^2 & \rho \\ \rho & (\sigma_2)^2 \end{bmatrix}$$

then the determinant

$$D \doteq |\mathbf{U}| = (\sigma_1)^2(\sigma_2)^2 - \rho^2 \tag{13}$$

---

9. We are using the $L(t)$ and $R(t)$ notation introduced in section 2.2.

If further $\mathbf{y} = (y_1, y_2)$ and $\mathbf{m} = (m_1, m_2)$ then we can rewrite (12) as

$$\mathcal{N}_i(\mathbf{y}) = \frac{1}{2\pi\sqrt{D(t)}} \exp\left\{-\frac{G(t)}{2D(t)}\right\} \tag{14}$$

where

$$G \doteq (\sigma_2)^2 (y_1 - m_1)^2 - 2\rho(y_1 - m_1)(y_2 - m_2) + (\sigma_1)^2 (y_2 - m_2)^2 \tag{15}$$

It follows that the parameter estimation task for real vector HMMs concerns the values of $\theta(t) \doteq \{p(t), m_1(t), m_2(t), \sigma_1(t)^2, \sigma_2(t)^2, \rho(t)\}$. Denoting by $\phi(t)$ any of the members of the set $\theta(t)$, we have shown in section 9.3 that we seek the solution to the equation[10]

$$\sum_t P_{\theta'}(\mathbf{t}|\mathbf{Y}) \frac{(\partial/\partial\phi(t)) P_\theta(\mathbf{t}, \mathbf{Y})}{P_\theta(\mathbf{t}, \mathbf{Y})} - \sum_m \lambda_m \sum_{t:L(t)=m} p_\theta(t) = 0 \tag{16}$$

$$\text{for all non-null } t \text{ and } \phi(t) \in \theta(t)$$

where θ' and θ denote the old and new (reestimated) parameter set values. Now[11]

$$P_\theta(\mathbf{t}, \mathbf{Y}) = \prod_{l=1}^{n} p(t_l) \prod_t \mathcal{N}_i(\mathbf{y}_l)^{\delta(t, t_l)} \tag{17}$$

where the observed sequence $\mathbf{Y}$ is of length n, and to simplify notation we have allowed no null transitions in the sequence $\mathbf{t}$. (Null transitions would make the indexing below more complex, but would change nothing of significance.)

First, observe that the transition probabilities $p(t)$ are not involved in the factor $\prod_t \mathcal{N}_i(\mathbf{y}_l)^{\delta(t, t_l)}$ at all; the latter is simply a constant that cancels out from the numerator and denominator of (16). Therefore, the estimation problem for $p(t)$ is essentially the same as that posed in (4), so that the reestimated $p_\theta(t)$ will be given by (compare with (8))

$$p_\theta(t) = \frac{1}{K} \sum_{l=1}^{n} \alpha_{l-1}^{\theta'}(L(t)) p_{\theta'}(t) \mathcal{N}_t^{\theta'}(\mathbf{y}_l) \beta_l^{\theta'}(R(t)) \tag{18}$$

where K is a normalizing constant, and α and β were defined in (9).

---

10. We are using the sequence notation $\mathbf{Y} = \mathbf{y}_1, \mathbf{y}_2, \ldots, \mathbf{y}_n$ where $\mathbf{y}_l = (y_{l,1}, y_{l,2})$.

11. Of course, $\prod_t \mathcal{N}_i(\mathbf{y}_l)^{\delta(t, t_l)} = \mathcal{N}_{t_l}(\mathbf{y}_l)$. We use the seemingly more complex product to be able to differentiate later with respect to functions of the transition variable t.

We will next take partial derivatives with respect to the means m_i. We have

$$\frac{\partial}{\partial m_i} \mathcal{N}_t(\mathbf{y}) = \frac{\partial}{\partial m_i} \left(\frac{1}{2\pi\sqrt{D}} \exp\left\{ -\frac{G}{2D} \right\} \right)$$

$$= \left(\frac{1}{2\pi\sqrt{D}} \exp\left\{ -\frac{G}{2D} \right\} \right) \left(-\frac{(\partial/\partial m_i)G}{2D} \right)$$

and

$$\frac{\partial}{\partial m_i} G = -2(\sigma_i)^2 (y_{l,i} - m_i) + 2\rho(y_{l,j} - m_j) \qquad j \neq i$$

so

$$\frac{\partial}{\partial m_i} \mathcal{N}_t(\mathbf{y}) = \left(\frac{1}{2\pi\sqrt{D}} \exp\left\{ -\frac{G}{2D} \right\} \right) \left(\frac{(\sigma_i)^2 (y_{l,i} - m_i) - \rho(y_{l,j} - m_j)}{D} \right)$$

It then follows directly from (14) and (17) that

$$\frac{(\partial)/\partial m_i(t'))P_\theta(\mathbf{t}, \mathbf{y})}{P_\theta(\mathbf{t}, \mathbf{y})}$$

$$= \sum_{l=1}^{n} \frac{(\partial/\partial m_i(t'))\mathcal{N}_{t'}(\mathbf{y}_l)^{\delta(t', t_l)}}{\mathcal{N}_{t'}(\mathbf{y}_l)^{\delta(t', t_l)}} \tag{19}$$

$$= \frac{1}{D(t')} \sum_{l=1}^{n} \delta(t', t_l)(\sigma_i(t')^2 (y_{l,i} - m_i(t')) - \rho(t')(y_{l,j} - m_j(t'))) \qquad j \neq i$$

We next want to derive the expressions for the partial derivatives with respect to σ_i and ρ. Denoting either by φ, we get

$$\frac{\partial}{\partial \varphi} \mathcal{N}_t(\mathbf{y}) = \frac{\partial}{\partial \varphi} \left(\frac{1}{2\pi\sqrt{D}} \exp\left\{ -\frac{G}{2D} \right\} \right)$$

$$= \left(\frac{1}{2\pi\sqrt{D}} \exp\left\{ -\frac{G}{2D} \right\} \right) \left(-\frac{1}{2D^2} \left((D - G)\frac{\partial}{\partial \varphi} D + D \frac{\partial}{\partial \varphi} G \right) \right) \tag{20}$$

so that

$$\frac{(\partial/\partial \varphi(t'))P_\theta(\mathbf{t}, \mathbf{y})}{P_\theta(\mathbf{t}, \mathbf{y})} = \sum_{l=1}^{n} \frac{(\partial/\partial \varphi(t'))\mathcal{N}_{t'}(\mathbf{y}_l)^{\delta(t', t_l)}}{\mathcal{N}_{t'}(\mathbf{y}_l)^{\delta(t', t_l)}}$$

$$= \sum_{l=1}^{n} \delta(t', t_l) \left(-\frac{1}{2D^2} \left((D - G)\frac{\partial}{\partial \varphi} D + D \frac{\partial}{\partial \varphi} G \right) \right)$$

Since

$$\frac{\partial}{\partial \varphi} D = \begin{cases} 2\sigma_i(\sigma_j)^2 & j \neq i \quad \text{if } \varphi = \sigma_i \\ -2p & \text{if } \varphi = p \end{cases} \tag{21}$$

$$\frac{\partial}{\partial \varphi} G = \begin{cases} 2\sigma_i(y_{l,j} - m_j)^2 & j \neq i \quad \text{if } \varphi = \sigma_i \\ -2(y_{l,1} - m_1)(y_{l,2} - m_2) & \text{if } \varphi = p \end{cases} \tag{22}$$

then we finally get

$$\frac{(\partial/\partial\sigma_i(t'))P_\theta(\mathbf{t}, \mathbf{y})}{P_\theta(\mathbf{t}, \mathbf{y})}$$

$$= \sum_{l=1}^{n} \delta(t', t_l) \frac{-2\sigma_i}{2((\sigma_1)^2(\sigma_2)^2 - p^2)^2}$$

$$\times \left[(\sigma_j)^2 \left((\sigma_1)^2(\sigma_2)^2 - p^2 - (\sigma_2)^2(y_{l,1} - m_1)^2 \right. \right.$$

$$\left. + 2p(y_{l,1} - m_1)(y_{l,2} - m_2) - (\sigma_1)^2(y_{l,2} - m_2)^2 \right)$$

$$\left. + ((\sigma_1)^2(\sigma_2)^2 - p^2)(y_{l,j} - m_j)^2 \right]$$

where we have omitted on the right-hand side the argument t' to keep the notation as simple as possible. Therefore

$$\frac{(\partial/\partial\sigma_1(t'))P_\theta(\mathbf{t}, \mathbf{y})}{P_\theta(\mathbf{t}, \mathbf{y})}$$

$$= \frac{-\sigma_1}{((\sigma_1)^2(\sigma_2)^2 - p^2)^2}$$

$$\times \sum_{l=1}^{n} \delta(t', t_l) \left((\sigma_1)^2(\sigma_2)^4 - p^2(\sigma_2)^2 - (\sigma_2)^4(y_{l,1} - m_1)^2 \right.$$

$$\left. + 2p(\sigma_2)^2(y_{l,1} - m_1)(y_{l,2} - m_2) - p^2(y_{l,2} - m_2)^2 \right) \tag{23}$$

$$= \frac{-\sigma_1}{((\sigma_1)^2(\sigma_2)^2 - p^2)^2}$$

$$\times \sum_{l=1}^{n} \delta(t', t_l) \left((\sigma_2)^4[(\sigma_1)^2 - (y_{l,1} - m_1)^2] \right.$$

$$- 2p(\sigma_2)^2[p - (y_{l,1} - m_1)(y_{l,2} - m_2)]$$

$$\left. + p^2[(\sigma_2)^2 - (y_{l,2} - m_2)^2] \right)$$

and

$$\frac{(\partial/\partial\sigma_2(t'))P_\theta(\mathbf{t},\mathbf{y})}{P_\theta(\mathbf{t},\mathbf{y})}$$

$$= \frac{-\sigma_2}{((\sigma_1)^2(\sigma_2)^2 - \rho^2)^2}$$

$$\times \sum_{l=1}^{n} \delta(t',t_l)\Big((\sigma_1)^4[(\sigma_2)^2 - (y_{l,2} - m_2)^2] \\ - 2\rho(\sigma_1)^2[\rho - (y_{l,1} - m_1)(y_{l,2} - m_2)] \\ + \rho^2[(\sigma_1)^2 - (y_{l,1} - m_1)^2]\Big)$$

(24)

Finally

$$\frac{(\partial/\partial\rho(t'))P_\theta(\mathbf{t},\mathbf{y})}{P_\theta(\mathbf{t},\mathbf{y})}$$

$$= \frac{-1}{2((\sigma_1)^2(\sigma_2)^2 - \rho^2)^2}$$

$$\times \sum_{l=1}^{n} \delta(t',t_l)\Big\{(-2\rho)\Big[(\sigma_1)^2(\sigma_2)^2 - \rho^2 \\ - \big((\sigma_2)^2(y_{l,1} - m_1)^2 - 2\rho(y_{l,1} - m_1)(y_{l,2} - m_2) \\ + (\sigma_1)^2(y_{l,2} - m_2)^2\big)\Big] \\ + ((\sigma_1)^2(\sigma_2)^2 - \rho^2)(-2(y_{l,1} - m_1)(y_{l,2} - m_2))\Big\}$$

$$= \frac{1}{((\sigma_1)^2(\sigma_2)^2 - \rho^2)^2}$$

$$\times \sum_{l=1}^{n} \delta(t',t_l)\Big[\rho((\sigma_1)^2(\sigma_2)^2 - \rho^2) \\ - \big(\rho(\sigma_2)^2(y_{l,1} - m_1)^2 \\ - ((\sigma_1)^2(\sigma_2)^2 + \rho^2)(y_{l,1} - m_1)(y_{l,2} - m_2) \\ + \rho(\sigma_1)^2(y_{l,2} - m_2)^2\big)\Big]$$

$$= \frac{-1}{((\sigma_1)^2(\sigma_2)^2 - \rho^2)^2}$$

$$\times \sum_{l=1}^{n} \delta(t',t_l)\Big[((\sigma_1)^2(\sigma_2)^2 + \rho^2)(\rho - (y_{l,1} - m_1)(y_{l,2} - m_2)) \\ + \rho(\sigma_2)^2((\sigma_1)^2 - (y_{l,1} - m_1)^2) \\ + \rho(\sigma_1)^2((\sigma_2)^2 - (y_{l,2} - m_2)^2)\Big]$$

It follows directly from (16) and the right-hand sides of (19), (23), (24), and (25) that the desired parameter setting $\theta(t')$ is one satisfying the following equations:[12]

$$\sum_t P_{\theta'}(\mathbf{t}, \mathbf{y}) \sum_{l=1}^{n} \delta(t', t_l)(y_{l,i} - m_i(t')) = 0 \quad i = 1, 2 \tag{26}$$

$$\sum_t P_{\theta'}(\mathbf{t}, \mathbf{y}) \sum_{l=1}^{n} \delta(t', t_l)[(y_{l,i} - m_i(t'))^2 - \sigma_i(t')^2] = 0 \quad i = 1, 2 \tag{27}$$

$$\sum_t P_{\theta'}(\mathbf{t}, \mathbf{y}) \sum_{l=1}^{n} \delta(t', t_l)((y_{l,1} - m_1(t'))(y_{l,2} - m_2(t')) - \rho(t')) = 0 \tag{28}$$

Thus (repeating (8) for convenience) the desired reestimation formulas are

$$p_\theta(t) = \frac{1}{K} \sum_{l=1}^{n} \alpha_{l-1}^{\theta'}(L(t)) \, p_{\theta'}(t) \mathcal{N}_t^{\theta'}(\mathbf{y}_l) \beta_l^{\theta'}(R(t)) \tag{29}$$

$$m_i^\theta(t) = \frac{1}{K^*(t)} \sum_{l=1}^{n} \alpha_{l-1}^{\theta'}(L(t)) \, p_{\theta'}(t) \mathcal{N}_t^{\theta'}(\mathbf{y}_l) \beta_l^{\theta'}(R(t)) \times y_{l,i} \quad i = 1, 2 \tag{30}$$

$$\sigma_i^\theta(t)^2 = \frac{1}{K^*(t)} \sum_{l=1}^{n} \alpha_{l-1}^{\theta'}(L(t)) \, p_{\theta'}(t) \mathcal{N}_t^{\theta'}(\mathbf{y}_l) \beta_l^{\theta'}(R(t))$$
$$\times (y_{l,i} - m_i^\theta(t))^2 \quad i = 1, 2 \tag{31}$$

$$\rho^\theta(t') = \frac{1}{K^*(t)} \sum_{l=1}^{n} \alpha_{l-1}^{\theta'}(L(t)) \, p_{\theta'}(t) \mathcal{N}_t^{\theta'}(\mathbf{y}_l) \beta_l^{\theta'}(R(t))$$
$$\times (y_{l,1} - m_1^\theta(t))(y_{l,2} - m_2^\theta(t)) \tag{32}$$

In (29) the normalizing constant K assures that the sum of the probabilities of transitions leaving any state equals 1. The normalizing function $K^*(t)$ is given by

$$K^*(t) = \sum_{l=1}^{n} \alpha_{l-1}^{\theta'}(L(t)) \, p_{\theta'}(t) \mathcal{N}_t^{\theta'}(\mathbf{y}_l) \beta_l^{\theta'}(R(t))$$

As always, we provide for each parameter to be estimated a counter (or a set of counters) that accumulates the terms of the above sums. This

12. In which, for convenience of actual computation, we have replaced $P_{\theta'}(\mathbf{t}|\mathbf{y})$ by $P_{\theta'}(\mathbf{t}, \mathbf{y})$.

is somewhat complicated in the case of formulas (31) and (32), which involve the results $m_i^\theta(t)$ of the computation of (30).

9.4.2   The Generalization to k Dimensions

It is obvious both from the above derivation and from its results (29) through (32) how to estimate parameters in the k-dimensional general case. In fact, if $\mathbf{U}(t) = [\rho_{i,j}(t)]$ denotes the covariance $k \times k$ matrix then the general reestimation formulas are simply (29), (30), and

$$\rho_{i,j}^\theta(t') = \frac{1}{K^*(t)} \sum_{l=1}^{n} \alpha_{l-1}^{\theta'}(L(t))\, p_{\theta'}(t)\, \mathcal{N}_t^{\theta'}(\mathbf{y}_l), \beta_l^{\theta'}(R(t))$$

$$\times\, (y_{l,i} - m_i^\theta(t))(y_{l,j} - m_j^\theta(t)) \tag{33}$$

9.5   Constant and Tied Parameters

9.5.1   Keeping Some Parameters Constant

In many applications, we wish to estimate the values of only some HMM parameters. For instance, in smoothing trigram models (see section 4.4) we wanted to find the probabilities λ_i of the model's null transitions while keeping the output distributions $f(\ |\ )$ as determined from the main part of the text training data. It is obvious from our derivation of the Baum-Welch algorithm for both symbolic (section 9.3) and real vector (section 9.4) outputs that

1. The formulas estimating the transition probabilities out of one state do not involve explicitly the transition probabilities out of another state. (Of course, the totality of all parameters affects the values of $\alpha_l(s)$ and $\beta_l(s)$ at various states s.)

2. The formulas estimating the output distributions or output density parameters associated with one transition do not involve explicitly the corresponding parameters associated with other transitions.

Thus the sole difference between estimating all the parameters and only *some* of them is that the latter case involves fewer counters (but of the same type).

In the case of real vectors, it is even possible to keep constant some means and covariances associated with a particular transition and estimate that transition's remaining parameters. Inspection of (19), (23), (24), and (25) shows what must be done with the contents of the counters in that case (to which (29), (30), and (33) no longer apply).

In current practice, the most common parameters to be kept constant are the off-diagonal covariances. Because of lack of sufficient data and the computational expense of their estimation, these covariances are usually set to equal 0 (i.e., $p_{i,j}(t) = 0$ for $i \neq j$). For this special case (29), (30), and (33) remain applicable.

9.5.2 Tying of Parameter Sets

In the vast majority of cases of interest, the HMMs used are so complex (and there are so many of them) that there is insufficient data to estimate the values of all their parameters individually. Neither does there exist enough outside knowledge to fix the values of certain parameters and estimate the others. The designer invariably partitions the set of parameters into subsets and insists that all the parameters belonging to the same subset have the same value,[13] which he sets out to estimate. The parameters of any given subset are then said to be *tied*.

Let us first see what the consequences of tying are in the case of the symbolic output alphabet treated in section 9.3. Suppose the transition probabilities out of states i and i' are to be tied in such a way that $p_{ij} = p_{i'j}$ for $j = 1, 2, \ldots, l_i$. Then, using the notation of the development of (3) through (7), the relation (5) becomes

$$\frac{(\partial/\partial p_{ij})P_\theta(\mathbf{t}, \mathbf{y})}{P_\theta(\mathbf{t}, \mathbf{y})} = \frac{c_{ij}(\mathbf{t}) + c_{i'j}(\mathbf{t})}{p_{ij}}$$

and therefore (compare with (7))

$$p_{ij} = \frac{1}{K_i} \sum_t P_{\theta'}(\mathbf{t}, \mathbf{y})[c_{ij}(\mathbf{t}) + c_{i'j}(\mathbf{t})]$$

The treatment of the general case is obvious. If the transition probabilities out of states $i, i', \ldots$ are to be tied, we simply pool the contents of their corresponding counters and after each iteration set the new probabilities to be proportional to the pool's final contents.[14]

---

13. The membership of these tied subsets is either determined intuitively or from data, usually by the method of *decision trees* discussed in chapter 10.

14. Note that this approach implies that the tied parameters are truly identical, that is, interchangeable. So the fact that the statistics for transition t may be derived mainly from observations of transition t' is presumably not bothersome. We would undoubtedly much rather not have the parameters identical but use one as a help in deriving a more robust estimate of the other. This is not what tying accomplishes.

We next focus our attention on the case of real vector outputs. Then tying of output parameters associated with two transitions t' and t'' means that $\mathbf{m}(t') = \mathbf{m}(t'')$ and $\mathbf{U}(t') = \mathbf{U}(t'')$. Therefore, $\theta^* \doteq \theta(t') = \theta(t'')$. (See the notation introduced following (15)). Again, let ϕ^* denote any of the members of the parameter set θ^*.

Using formula (17) we get

$$
\frac{(\partial/\partial\phi^*)P_\theta(\mathbf{t}, \mathbf{y})}{P_\theta(\mathbf{t}, \mathbf{y})} = \frac{(\partial/\partial\phi^*)\prod_{l=1}^{n}p(t_l)\prod_t \mathcal{N}_t(\mathbf{y}_l)^{\delta(t,t_l)}}{\prod_{l=1}^{n}p(t_l)\prod_t \mathcal{N}_t(\mathbf{y}_l)^{\delta(t,t_l)}}
$$

$$
= \sum_{l=1}^{n}\left(\frac{(\partial/\partial\phi^*)\mathcal{N}_{t'}(\mathbf{y}_l)^{\delta(t',t_l)}}{\mathcal{N}_{t'}(\mathbf{y}_l)^{\delta(t',t_l)}} + \frac{(\partial/\partial\phi^*)\mathcal{N}_{t''}(\mathbf{y}_l)^{\delta(t'',t_l)}}{\mathcal{N}_{t''}(\mathbf{y}_l)^{\delta(t'',t_l)}}\right)
$$

$$
= \sum_{l=1}^{n}(\delta(t', t_l) + \delta(t'', t_l))\frac{(\partial/\partial\phi^*)\mathcal{N}_{t_l}(\mathbf{y}_l)}{\mathcal{N}_{t_l}(\mathbf{y}_l)}
$$

As a result, condition (26) becomes

$$
\sum_t P_{\theta'}(\mathbf{t}|\mathbf{y}) \sum_{l=1}^{n}(\delta(t', t_l) + \delta(t'', t_l))(y_{l,i} - m_i(t_l)) = 0 \qquad i = 1, 2
$$

and so (30) is replaced by

$$
m_i^\theta(t') = m_i^\theta(t'')
$$

$$
= \frac{1}{K^*(t') + K^*(t'')} \sum_{l=1}^{n}\left(\alpha_{l-1}^{\theta'}(L(t'))\, p_{\theta'}(t')\mathcal{N}_{t'}^{\theta'}(\mathbf{y}_l)\, \beta_l^{\theta'}(R(t'))\right.
$$

$$
\left. + \alpha_{l-1}^{\theta'}(L(t''))\, p_{\theta'}(t'')\mathcal{N}_{t''}^{\theta'}(\mathbf{y}_l)\, \beta_l^{\theta'}(R(t''))\right) \times y_{l,i}
$$

$$
i = 1, 2
$$

and similarly (31) is replaced by

$$
\sigma_i^\theta(t')^2 = \sigma_i^\theta(t'')^2
$$

$$
= \frac{1}{k^*(t') + k^*(t'')} \sum_{l=1}^{n}\left(\alpha_{l-1}^{\theta'}(L(t'))\, p_{\theta'}(t')\mathcal{N}_{t'}^{\theta'}(\mathbf{y}_l)\, \beta_l^{\theta'}(R(t'))\right.
$$

$$
\left. + \alpha_{l-1}^{\theta'}(L(t''))\, p_{\theta'}(t'')\mathcal{N}_{t''}^{\theta'}(\mathbf{y}_l)\, \beta_l^{\theta'}(R(t''))\right) \times (y_{l,i} - m_i^\theta(t'))^2
$$

and (32) by

$$\rho^\theta(t') = \rho^\theta(t'')$$

$$= \frac{1}{k^*(t') + k^*(t'')} \sum_{l=1}^{n} \left(\alpha_{l-1}^{\theta'}(L(t')) p_{\theta'}(t') \mathcal{N}_{t'}^{\theta'}(\mathbf{y}_l) \beta_l^{\theta'}(R(t'))) \right.$$

$$\left. + \alpha_{l-1}^{\theta'}(L(t'')) p_{\theta'}(t'') \mathcal{N}_{t''}^{\theta'}(\mathbf{y}_l) \beta_l^{\theta'}(R(t''))) \right)$$

$$\times (y_{l,1} - m_1^\theta(t'))(y_{l,2} - m_2^\theta(t'))$$

We thus see that all that needs to be done, even for estimating real vector output parameters, is simply the pooling of contents of counters corresponding to those transitions whose output probabilities are to be tied.

9.6   Tied Mixtures

The current continuous speech recognizers use as acoustic processor outputs vectors of cepstral coefficients [3]. Their production probabilities must be tied not only because there is insufficient data to estimate them, but also because the calculation at recognition time of candidate probabilities pertaining to the observed vectors $\mathbf{y}_l$ may be computationally too expensive.

The ingenious accepted remedy is referred to as the method of *tied mixtures* [4]. The number of different normal output densities $\mathcal{N}_i(\mathbf{y})$, $i = 1, 2, \ldots, M$ is limited to some tolerable number M, and the output density corresponding to any particular transition t is specified by the formula

$$P(\mathbf{y}|t) = \sum_{i=1}^{M} q(i|t) \mathcal{N}_i(\mathbf{y}) \tag{34}$$

The computational saving is obvious. As the acoustic processor generates the successive output vectors $\mathbf{y}_l$, the recognizer computes (possibly in parallel) the values $\mathcal{N}_i(\mathbf{y})$, $i = 1, 2, \ldots, M$. These values are then employed in formulas (34) when evaluating the likelihood of transitions through various HMMs. Of course, there is no requirement that for any particular i, t combination, $q(i|t) > 0$. So although M may even be of the order of several hundred, the number of nonzero terms in (34) would typically be of the order of ten.

There is an additional important reason for employing the mixture formula. Although the Baum-Welch algorithm conveniently estimates the

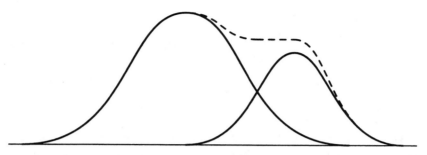

Figure 9.1
A density function resulting from the superposition of two Gaussian densities

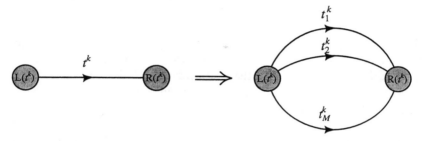

Figure 9.2
Replacement of a single transition by a set illustrating an analysis of probability
weight estimation for tied mixture distributions

parameters of normal densities, these do not adequately reflect the speech
process. Because of their squared exponent, *Gaussian* density values *decay*
too fast as the observable variable deviates from its mean. The simple
expedient of mixing several densities having different means can slow this
rate of decay. Figure 9.1 illustrates this for the one-dimensional case.

In principle, it is easy to estimate the parameters of HMMs whose
outputs are generated according to the formula (34). We need only to
replace in the HMM a single transition t^k by M parallel transitions
$t_i^k, i = 1, 2, \ldots, M$ having the same source and target states $L(t^k)$ and
$R(t^k)$.[15] The output density associated with the transition t_i^k is then $\mathcal{N}_i(\mathbf{y})$.
During the parameter estimation process all the output densities for dif-

15. See figure 9.2.

ferent values of k at transitions t_i^k are tied while the transition probabilities $p(t_{i,}^k) = q(i|t^k)$ are determined in the usual way by formula (29).

9.7 Additional Reading

An easy to read exposition of the EM algorithm with some accompanying examples can be found in reference [5]. Since the EM algorithm is in many cases computationally expensive, the possibilities of acceleration [6] as well as the question of the rate of convergence [7] are of overriding interest.

We pointed out in section 9.6 the usefulness of tied mixtures [4]. Independent mixtures, however, give potentially even better results provided enough training material is available for accurate estimation [8]–[12]. State-of-the-art methods of tying [13] are based on growing decision trees as described in chapter 10. We will return to the specific problem of HMM state tying in chapter 12.

References

[1] A.P. Dempster, N.M. Laird, and D.B. Rubin, "Maximum likelihood from incomplete data via the EM algorithm", *Journal of the Royal Statistical Society*, ser. B, vol. 39, pp. 1–38, 1977.

[2] L. Baum, "An inequality and associated maximization technique in statistical estimation of probabilistic functions of a Markov process," *Inequalities*, vol. 3, pp. 1–8, 1972.

[3] S.B. Davis and P. Mermelstein, "Comparison of parametric representations for monosyllabic word recognition in continuously spoken sentences," *IEEE Transactions on Acoustics, Speech, and Signal Processing*, vol. ASSP-28, no. 4, pp. 357–66, August 1980.

[4] J.R. Bellegarda and D. Nahamoo, "Tied mixture continuous parameter modeling for speech recognition," *IEEE Transactions on Acoustics, Speech, and Signal Processing*, vol. ASSP-38, no, 12, pp. 2033–45, December 1990.

[5] T.K. Moon, "The Expectation-Maximization algorithm," *IEEE Signal Processing Magazine*, vol. 13, no. 6, pp. 47–60, November 1996.

[6] Isaac Meilijson, "A fast improvement to the EM algorithm on its own terms," *Journal of Royal Statistical Society*, ser. B, vol. 51, no. 1, pp. 127–38, 1989.

[7] C.F.J. Wu, "On the convergence properties of the EM algorithm," *The Annals of Statistics*, vol. 11, no. 1, pp. 95–103, 1983.

[8] A. Nadas and D. Nahamoo, "Automatic speech recognition via pseudo-independent marginal mixtures," *Proceedings of the IEEE International Conference on Acoustics, Speech, and Signal Processing*, pp. 1285–87, Dallas, Tx, April 1987.

[9] R.A. Redner and H.F. Walker, "Mixture densities, maximum likelihood and the EM algorithm," *SIAM Review*, vol. 26, no. 2, pp. 195–39, April 1984.

[10] L.R. Rabiner, B.H. Juang, S.E. Levinson, and M.M. Sondhi, "Recognition of isolated digits using hidden Markov models with continouous mixture densities," *AT&T Technical Journal*, vol. 64, no. 6, pp. 1211–34, July-August 1985.

[11] B.H Juang, "Maximum-likelihood estimation for mixture multivariate stochastic observations of Markov chains," *AT&T Technical Journal*, vol. 64, no. 6, pp. 1235–49, July-August 1985.

[12] L.R. Rabiner, B.H. Juang, S.E. Levinson, and M.M. Sondhi, "Some properties of continuous hidden Markov model representations," *AT&T Technical Journal*, vol. 64, no. 6, pp. 1251–70, July-August 1985.

[13] S.J. Young, J.J. Odell, P.C. Woodland, "Tree-based state tying for high accuracy acoustic modelling," *Proceedings of the DARPA Speech and Natural Language Processing Workshop*, 307–12, Plainsboro, March 1994.

Chapter 10
Decision Trees and Tree Language Models

10.1 Introduction

This chapter is dedicated to *decision trees*, a very powerful technique of equivalence classification based on training data. Because we always want to be concrete, we will discuss it in the context of language modeling, which in many ways provides the simplest and at the same time most general setting. The technique was developed in statistics [1] and has been successfully applied to many aspects of the speech recognition problem. Two such applications will be presented in the following chapters.

Although decision trees are very attractive in principle, they suffer certain fundamental flaws that the current state of the art treats in a rather ad hoc manner. Decision trees are a greedy method of iterative refinement of equivalence classes. Experimental investigations seem to show, however, that greed is only a minor problem. Decision trees have two major flaws: training data fragmentation (see section 10.14.2) and the absence of a theoretically founded stopping criterion.[1]

We will first show that decision tree construction is equivalent to successive refinement of equivalence classes driven by answers to questions. We will introduce cross-validation [2] as a good basis for a stopping criterion (section 10.5). We will then describe Chou's method [3] of quasi-optimal question determination (section 10.7). Finally, we will describe another decision tree construction method, due to Mercer [4], specifically aimed at language modeling (section 10.11). Throughout this chapter we will pay attention to practical considerations and will point out various shortcomings of the current state of the art (section 10.10, 10.13, and 10.14).

---

1. That is, a criterion to stop the further refinement of the equivalence classes developed up to that point. See section 10.5.

Mercer's method is based on a very interesting word clustering process capable of automatic discovery of word (semantic/syntactic) equivalence classes (section 10.12). It is important in its own right and can be thought of as a significant by-product of decision tree language modeling.

10.2   Application of Decision Trees to Language Modeling

A large vocabulary speech recognizer aims to find that word sequence $\hat{\mathbf{W}}$ satisfying the relation

$$\hat{\mathbf{W}} = \arg \max_{\mathbf{W}} P(\mathbf{A}|\mathbf{W})P(\mathbf{W})$$

where $\mathbf{A}$ denotes the string of acoustic observations and $\mathbf{W} = w_1, w_2, \ldots$, a string of words. A language model is a device that provides for every word string $\mathbf{W}$ the probability $P(\mathbf{W})$ that the speaker will utter $\mathbf{W}$. In chapter 4 we discussed the widely used trigram language model, which computes $P(\mathbf{W})$ by the formula

$$P(\mathbf{W}) = \prod_i P(w_i|w_{i-2}, w_{i-1})$$

Let $\mathbf{h}_i = w_1, w_2, \ldots, w_{i-1}$ denote the word generation history before the ith word is produced. As we pointed out in chapter 4, language modeling's general problem is to find a history equivalence classifier $\Phi(\mathbf{h})$ that can be used to estimate the probabilities $P(w|\Phi(\mathbf{h}))$ so as to make the calculation $P(\mathbf{W}) = \prod_i P(w_i|\Phi(\mathbf{h}_i))$ the most effective basis possible for correct recognition. In chapter 8 we argued that a good measure of such effectiveness is the entropy of the word production process induced by the probability $P(\mathbf{W})$. The decision tree method uses entropy as a criterion in developing the requisite equivalence classification Φ and thereby the probability estimates $P(w|\Phi(\mathbf{h}))$.

10.3   Decision Tree Example

In a tree language model successive answers to questions about the history $\mathbf{h}$ determine the equivalence class $\Phi(\mathbf{h})$ [5]. For instance, in the example of figure 10.1, the classifier first asks question Q_{10} about $\mathbf{h}$. If the answer is *No* then question Q_{21} is posed. If the answer to Q_{21} is *Yes*, then Q_{32} is posed. If this answer is *Yes*, then $\mathbf{h}$ is classified as belonging to class Φ_{44}; if the answer is *No* then $\mathbf{h}$ belongs to class Φ_{45}. Note that the tree in

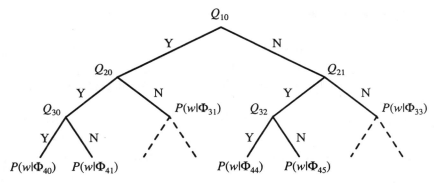

Figure 10.1
Equivalence classification induced on a tree

figure 10.1 is not uniformly developed. So if Q_{20} is answered *No* then **h** will be assigned to Φ_{31}, and no further questions will be asked.

How would one estimate the probabilities $P(w|\Phi_{lj})$ for the equivalence classes in the tree of figure 10.1?[2] The simplest method would be to create from a text training corpus word collections $\mathscr{S}_{lj}$ corresponding to tree leaves Φ_{lj}.[3] These collections $\mathscr{S}_{lj}$ will consist of those words w_i that followed in that corpus such histories $\mathbf{h}_i$, $i = 1, \ldots, n$, for which $\Phi(\mathbf{h}_i) = \Phi_{lj}$. The probability $P(w|\Phi_{lj})$ would then be determined according to the number of times the word w was found in the collection $\mathscr{S}_{lj}$.[4]

We close this section by pointing out that any equivalence classification $\Psi(\mathbf{h})$ can be encoded in a tree. Let $\Psi_1, \Psi_2, \ldots, \Psi_M$ be the complete partitioning of the set of histories. Then, for instance, we can let $\Phi_{20} = \bigcup_{j=1}^{K} \Psi_j, \Phi_{21} = \bigcup_{j=K+1}^{M} \Psi_j$, and $\Phi_{30} = \bigcup_{j=1}^{L} \Psi_j, \Phi_{31} = \bigcup_{j=L+1}^{K} \Psi_j$, etc. So obviously if we could find optimal questions, decision trees would lead to an optimal equivalence classification and thus to an optimal language model. Thus restricting equivalence classes to those derived from a decision tree causes by itself no loss of generality.

---

2. That is, once the questions attached to the decision tree's internal nodes are determined.

3. In figure 10.1, the leaf equivalence classes are Φ_{40}, Φ_{41}, Φ_{31}, Φ_{44}, Φ_{45}, and Φ_{33}.

4. In reality, $P(w|\Phi_{lj})$ would be equal to an appropriately smoothed version of the relative frequency $f(w|\Phi_{lj})$ determined from the collection $\mathscr{S}_{lj}$. Section 10.14.1 discusses how to carry out such smoothing.

10.4 What Questions?

In designing decision trees we face three problems:

1. What types of questions should be asked of the history?
2. What actual questions should be asked at a particular stage of tree development?
3. Which question selection criterion should be used in the tree construction process?

In section 10.2 we already disposed of the third problem: We chose entropy as the criterion.

The first problem is hardest: The possibilities seem infinite. Consider, for instance, the prediction of the three unknown words $w_1 w_2 w_3$ that are to follow the sentence fragment

Finite-state methods occupy a special position. . .

1. It would be reasonable to guess that w_1 is either "." (because the history h_1 is a possible complete sentence) or "in" (because that preposition is likely to follow the word "position," the object of the verb "occupy"). Hence a question like *Does h_1 possibly constitute a complete sentence?* may be appropriate.
2. Assume that $w_1 =$ "in." Then surely w_2 is either a determiner or a noun modifier (e.g., an adjective), and it is most likely "the" because of the frequency of the trigram "position in the." This suggests a question such as *Do the most likely parts of speech of the word following the last two words of h_2 belong to the (appropriately chosen) set $\mathscr{S}$?*
3. Assume that $w_1 w_2 =$ "in the." Then w_3 is probably an abstract noun, because it follows the determiner "the," and we are entitled to conclude that the word "position" is meant abstractly (and not concretely as in "the third position in the list is occupied by . . .") since it is a part of a complement of "occupy" in a sentence with the abstract subject "methods." So a good question might be *Is h_3 likely to be followed by an abstract noun?*

From this rather crude analysis we see how difficult, if not impossible, it would be to list the set of *all* permissible and intuitively satisfying questions from which the entropy criterion would choose the actual ones asked. Let us then postpone the problem to sections 10.6 and following and proceed to consider our remaining concern: the selection during tree development of actual questions from a list.

If we had a set of permissible questions as well as a goodness criterion[5] by which to evaluate them (i.e., a solution to problems one and three above), then we could solve problem two by brute force (more or less) via the following greedy approach (the notation is that of figure 10.1):

1. From among all possible questions, find that question Q_{10} that will classify all histories **h** into one of two equivalence classes Φ_{20} and Φ_{21} such that the resulting language model

$$P(\mathbf{W}) = \prod_i P(w_i|\Phi(\mathbf{h}_i))$$

is the best possible (according to the selected goodness criterion).

2. Keeping Q_{10} fixed, find the best tentative question Q_{20} that will help split the history equivalence class Φ_{20} into two classes Φ_{30} and Φ_{31} (if such a split were to be accomplished, histories **h** would at that stage belong to one of three equivalence classes: Φ_{30}, Φ_{31}, and Φ_{21}) such that the resulting language model is the best possible.

3. Keeping Q_{10} fixed, find the best tentative question Q_{21} that will help split Φ_{21} into two classes Φ_{32} and Φ_{33} (histories **h** would belong to one of three equivalence classes: Φ_{20}, Φ_{32}, and Φ_{33}) such that the resulting language model is the best possible.

4. Make permanent whichever of the two tentative questions Q_{20} or Q_{21} better satisfies the goodness criterion.

5. The pattern is now set. Assuming that Q_{21} was chosen, we will next choose tentative questions Q_{20}, Q_{32}, and Q_{33}, and make permanent the one that gives the best language model.

6. And so forth, continuing the pattern set.

The above process would go on until all leaves of the developed tree would correspond to unique histories found in the training data. This would result in overtraining.[6] So we will need a stopping criterion to guard us from such a calamity. This is a difficult problem that must be

---

5. We already suggested using empirical entropy, which can be computed by formula (1) in section 10.5.

6. Using the usual $\ln x \le x - 1$ inequality, it is easy to show that any question (i.e., any split) whatever will lower (or at least not increase) the entropy value based on the relative frequency of events in the training set (rather than on the actual probability distribution, which is, of course, unknowable). If the tree is further and further refined, the leaves end up referring to a unique past whose future they predict with certainty.

faced in all training situations and that so far has not yielded to any universally acceptable solution. A possible approach is outlined next.

10.5   The Entropy Goodness Criterion for the Selection of Questions, and a Stopping Rule

Let us divide the training text into a larger *development text* and a smaller *cross-validation check text*. As we construct a decision tree by applying the algorithm at the end of section 10.4 to the development text, we will construct simultaneously another tree that applies the same questions to the check text. We are thus constructing two trees in parallel. Which questions are selected depends on their effect on the goodness criterion applied to the development text.

The difference between the two trees will consist of the contents of the pairs of (two different) word sets $\mathscr{S}_{ij}$ corresponding to those leaves in the two trees specified by the same equivalence classes Φ_{ij}.[7] The respective word sets $\mathscr{S}_{ij}$ will be different, because they will be based on different data: those belonging to the development and check texts.[8] We will use the tree based on the check text to validate the hypothesis that the language model being developed is appropriate to as yet unseen data.[9] This approach is commonly referred to as *cross-validation* [2].

For the purposes of our goodness criterion, let us define the probabilities

$$P(w|\Phi_{ij}) = f(w|\mathscr{S}_{ij})$$

where f denotes the relative frequency of the word w in the corresponding word set,

$$f(w|\mathscr{S}_{ij}) = \frac{C(w|\mathscr{S}_{ij})}{C(\mathscr{S}_{ij})}$$

---

7. As stated in section 10.3, the word collection $\mathscr{S}_{ij}$ consists of those words w_i which in the corpus being considered follow such histories $\mathbf{h}_i, i = 1, \ldots, n$, for which $\Phi(\mathbf{h}_i) = \Phi_{ij}$.

8. The sets $\mathscr{S}_{ij}$ of next words that follow histories belonging to the same equivalence class Φ_{ij} will in general be different for the two texts.

9. There are many other possible stopping rules, most of which do not depend on validation data. In a very popular one, due to Chou [6], branch splitting continues as long as the amount of training data allocated to the branch exceeds some threshold, and then an algorithm is applied to prune the tree back.

Above, $C(\mathscr{S}_{ij})$ and $C(w|\mathscr{S}_{ij})$ denote the total number of words in $\mathscr{S}_{ij}$ (i.e., the size of the set $\mathscr{S}_{ij}$) and the number of occurrences of word w in the set $\mathscr{S}_{ij}$. Finally, let

$$f(\mathscr{S}_{ij}) = \frac{C(\mathscr{S}_{ij})}{n}$$

where n is the size of the (development or check) text.

The average empirical conditional entropy of the trees of either of the texts (where the conditioning is on the equivalence class of the history) is then given by

$$H \doteq - \sum_{i,j} \sum_{w} f(\mathscr{S}_{ij}) f(w|\mathscr{S}_{ij}) \log f(w|\mathscr{S}_{ij}) \tag{1}$$

where the sum is over all the leaves i,j of the tree. This entropy is the proposed goodness criterion.

We will use cross-validation to refine the tree construction algorithm of section 10.4. The evaluation on the check text will be based on the cross-entropy function

$$H^c \doteq - \sum_{i,j} \sum_{w} g(\mathscr{S}_{ij}) g(w|\mathscr{S}_{ij}) \log f(w|\mathscr{S}_{ij}) \tag{2}$$

where $f(w|\mathscr{S}_{ij})$ is the relative frequency evaluated over the development text and $g(\mathscr{S}_{ij})$ and $g(w|\mathscr{S}_{ij})$ are relative frequencies evaluated over the check text.

1. From among all possible questions, find that question Q_{10} that will classify all histories **h** into one of two equivalence classes Φ_{20} and Φ_{21} such that the resulting value of H computed on the development text is minimal. Call this value H_d. Using question Q_{10} evaluate H^c over the check text and call its value H_0^c.

2. Keeping Q_{10} fixed, find the best tentative question Q_{20} that when splitting Φ_{20} into two equivalence classes Φ_{30} and Φ_{31} will result in a minimal value of H (computed on the development text). Call this value H_{20}.

3. Keeping Q_{10} fixed, find the best tentative question Q_{21} which when splitting Φ_{21} into two equivalence classes Φ_{32} and Φ_{33} will result in a minimal value of H (computed on the development text). Call this value H_{21}.

4. Of Q_{20} or Q_{21}, eliminate the one that results in a larger value of H_{2i} ($i = 0, 1$). Let the retained question be Q_{2j}. If $H_d - H_{2j} < \varepsilon$, then stop

and reject Q_{2j} as well. Otherwise evaluate H_{2j}^c over the check text and call it H_1^c. If $H_0^c - H_1^C < \delta$ then stop. Else make question Q_{2j} permanent and set $H_d = H_{2j}$ and $H_0^c = H_1^c$.

5. The pattern is now set. Assuming that Q_{21} was chosen, we will next choose tentative questions Q_{20}, Q_{32}, and Q_{33}, and make permanent the one that lowers the entropy of the development text by at least ε and, simultaneously, conditional *cross-entropy* of the check text by at least δ. If both these improvements do not take place, no additional question will be accepted and the tree construction process will terminate.

6. And so forth, continuing the pattern set.

Once the tree is constructed, the relative frequencies based on the sets $\mathscr{S}_{ij}$ corresponding to the final leaves are recalculated from all the training data (i.e., from the combined development and check texts).

The stopping rule invoked in the above tree development algorithm is very conservative. It may lead to unnecessarily shallow trees. A less stringent rule would modify steps 4 and 5 as follows:

4'. Do for $j = 1, 2$:

If $H_d - H_{2j} < \varepsilon$ then reject Q_{2j} and bar leaf Φ_{2j} from ever being split. Otherwise evaluate H_{2j}^c over the check text. If $H_0^c - H_{2j}^c < \delta$ then bar leaf Φ_{2j} from ever being split. Else make question Q_{2j} permanent and set $H_d = H_{2j}$ and $H_0^c = H_{2j}^c$.

5'. The pattern is now set. At each successive depth of the tree, attempt to split its leaves in natural order. Bar leaves from further splitting if the entropy improvement made possible by the best question is insufficient when evaluated on either the development or the check tree. The process stops when no leaves are found that are allowed to be split.

10.6   A Restricted Set of Questions

In this section we make the task of tree language model construction easier and more concrete by insisting that the questions that can be asked of the history $\mathbf{h} = \ldots w_{-j}, w_{-j+1}, \ldots, w_{-2}, w_{-1}$ have the following form:

Is the word w_{-m} ($m = 1, 2, \ldots$) in history $\mathbf{h}$ a member of the set $\mathscr{S}$?[10]

10. This restriction obviously loses a lot of generality. Equivalence classes cannot now depend on particular characteristics of subsequences of the history. The approach of Bahl et al. [5], though also not fully general, does not give up as much.

The tree construction process then consists of

1. determining for each history position m the set $\mathcal{S}_m$ that would lead to the largest reduction in entropy if the above question were asked,
2. finding the minimizing $\hat{m}$, $\mathcal{S}_{\hat{m}}$ combination, $\hat{m} \in \{1, 2, \ldots\}$,
3. and finally, as a consequence, splitting the leaf being considered by the question *Is the word* $w_{-\hat{m}}$ *a member of the set* $\mathcal{S}_{\hat{m}}$?

The task then is to find the optimal sets $\mathcal{S}$ under the above restriction. If a limited list of set alternatives, specified in advance, is not provided, then a brute force search is not feasible. In fact, there are $2^{|\mathcal{V}|-1} - 1$ non–trivially different subsets $\mathcal{S}$ of elements from a given vocabulary $\mathcal{V}$. This is a huge choice even for moderate vocabulary sizes $|\mathcal{V}|$. In sections 10.7, 10.9, and 10.11 we will describe several algorithms for finding good subsets $\mathcal{S}$.

10.7 Selection of Questions by Chou's Method

In this section we present a powerful method due to Chou [3] that leads to a locally optimal split of a leaf. The local restriction is of the same kind as that of the Baum-Welch algorithm: Chou's iterative method improves the quality of the split after each iteration but is not guaranteed to find the global optimum. Like almost all other methods we have presented, this one is greedy: It is performed without any look ahead. The reader will surely note the similarity of Chou's method to k-means clustering (see Section 1.5). For concreteness, we will again use entropy as the goodness criterion, but the method applies to all convex criteria.[11]

Let $\mathcal{S}$ be the set of histories $\mathbf{h}$ belonging to the parent node we are trying to split, and let $f(w|\mathcal{S})$ be the relative frequency of the data to be predicted, as found at that node, and let $\beta_1, \beta_2, \ldots, \beta_N$ denote an *atomic partitioning* of $\mathcal{S}$, that is,

$$\beta_i \cap \beta_j = \phi \quad \text{for } i \neq j; \qquad \bigcup_{i=1}^{N} \beta_i = \mathcal{S} \tag{3}$$

Our intention is to split $\mathcal{S}$ into complementary subsets $\mathcal{A}$ and $\bar{\mathcal{A}}$ made up of elements of the atomic partition $\{\beta_1, \beta_2, \ldots, \beta_N\}$. We will write

$$f(w|\beta_i) \doteq f(w|\mathbf{h} \in \beta_i) \quad \text{and} \quad f(w|\mathcal{A}) \doteq f(w|\mathbf{h} \in \mathcal{A})$$

---

11. See [3]. A particularly interesting alternative is the Gini index discussed in section 10.10.1.

The idea of the atomic partition is that, for instance,

$$\beta_v = \{\mathbf{h} = w_{-1}, w_{-2}, \ldots : w_{-m} = v, v \in \mathscr{V}, \mathbf{h} \in \mathscr{S}\}$$

so that these atoms conform to the restrictions imposed in section 10.6.[12] We deal with unspecified (except for (3)) atoms to describe the algorithm in its generality.

We are seeking the split $\mathscr{A}$ for which

$$H_{\mathscr{A}} \doteq f(\mathscr{A})H(w|\mathscr{A}) + f(\bar{\mathscr{A}})H(w|\bar{\mathscr{A}}) \tag{4}$$

is minimum, where

$$H(w|\mathscr{A}) \doteq -\sum_w f(w|\mathscr{A})\log f(w|\mathscr{A}) \tag{5}$$

Defining

$$H_{\mathscr{S}} \doteq -\sum_w f(w|\mathscr{S})\log f(w|\mathscr{S})$$

$$H(w|\beta) \doteq -\sum_w f(w|\beta)\log f(w|\beta)$$

and

$$H_{\min} \doteq \sum_{i=1}^N f(\beta_i)H(w|\beta_i)$$

we see that minimizing $H_{\mathscr{A}}$ is the same as minimizing

$$
\begin{aligned}
H_{\mathscr{A}} &- H_{\min} \\
&= \sum_{\beta \in \mathscr{A}} f(\beta)[H(w|\mathscr{A}) - H(w|\beta)] + \sum_{\beta \in \bar{\mathscr{A}}} f(\beta)[H(w|\bar{\mathscr{A}}) - H(w|\beta)] \\
&= \sum_{\beta \in \mathscr{A}} \sum_w f(\beta)f(w|\beta)\log \frac{f(w|\beta)}{f(w|\mathscr{A})} + \sum_{\beta \in \bar{\mathscr{A}}} \sum_w f(\beta)f(w|\beta)\log \frac{f(w|\beta)}{f(w|\bar{\mathscr{A}})}
\end{aligned}
\tag{6}
$$

Suppose $\mathscr{S}$ is initially partitioned into sets $\mathscr{A}$ and $\bar{\mathscr{A}}$. Let us then repartition $\mathscr{S}$ into new complementary sets $\mathscr{A}^*$ and $\bar{\mathscr{A}}^*$ by the following rule:

---

12. Note that some atoms β_v defined by (3) may turn out to be empty because there may exist in the training data no history $\mathbf{h} \in \mathscr{S}$ such that $w_{-m} = v$. Note also that the values of m and $\mathscr{S}$ are considered fixed in (3).

place $\beta \in \mathscr{S}$ into $\mathscr{A}^*$ whenever

$$\left[\sum_w f(w|\beta)\log \frac{f(w|\beta)}{f(w|\mathscr{A})}\right] \le \left[\sum_w f(w|\beta)\log \frac{f(w|\beta)}{f(w|\bar{\mathscr{A}})}\right]$$

place $\beta \in \mathscr{S}$ into $\bar{\mathscr{A}}^*$ whenever $\qquad(7)$

$$\left[\sum_w f(w|\beta)\log \frac{f(w|\beta)}{f(w|\mathscr{A})}\right] > \left[\sum_w f(w|\beta)\log \frac{f(w|\beta)}{f(w|\bar{\mathscr{A}})}\right]$$

It then follows directly from the last line of (6) that

$$\begin{aligned}
H_{\mathscr{A}} - H_{\min} \ge & \sum_{\beta \in \mathscr{A}^*} \sum_w f(\beta)f(w|\beta)\log \frac{f(w|\beta)}{f(w|\mathscr{A})} \\
& + \sum_{\beta \in \bar{\mathscr{A}}^*} \sum_w f(\beta)f(w|\beta)\log \frac{f(w|\beta)}{f(w|\bar{\mathscr{A}})}
\end{aligned} \qquad(8)$$

Furthermore, it follows from the inequality $\ln x \le (x - 1)$ that

$$\begin{aligned}
\sum_{\beta \in \mathscr{A}^*} & \sum_w f(\beta)f(w|\beta)\log \frac{f(w|\mathscr{A})}{f(w|\mathscr{A}^*)} + \sum_{\beta \in \bar{\mathscr{A}}^*} \sum_w f(\beta)f(w|\beta)\log \frac{f(w|\bar{\mathscr{A}})}{f(w|\bar{\mathscr{A}}^*)} \\
= & \sum_w \sum_{\beta \in \mathscr{A}^*} f(\beta)f(w|\beta)\log \frac{f(w|\mathscr{A})}{f(w|\mathscr{A}^*)} + \sum_w \sum_{\beta \in \bar{\mathscr{A}}^*} f(\beta)f(w|\beta)\log \frac{f(w|\bar{\mathscr{A}})}{f(w|\bar{\mathscr{A}}^*)} \\
= & \sum_w f(\mathscr{A}^*)f(w|\mathscr{A}^*)\log \frac{f(w|\mathscr{A})}{f(w|\mathscr{A}^*)} + \sum_w f(\bar{\mathscr{A}}^*)f(w|\bar{\mathscr{A}}^*)\log \frac{f(w|\bar{\mathscr{A}})}{f(w|\bar{\mathscr{A}}^*)} \\
\le & f(\mathscr{A}^*)\left[\sum_w f(w|\mathscr{A}) - 1\right] + f(\bar{\mathscr{A}}^*)\left[\sum_w f(w|\bar{\mathscr{A}}) - 1\right] = 0 \qquad(9)
\end{aligned}$$

Adding the left-hand side of (9) (a non-positive number) to the right-hand side of (8) we then get

$$\begin{aligned}
H_{\mathscr{A}} - H_{\min} \ge & \sum_{\beta \in \mathscr{A}^*} \sum_w f(\beta)f(w|\beta)\log \frac{f(w|\beta)}{f(w|\mathscr{A}^*)} \\
& + \sum_{\beta \in \bar{\mathscr{A}}^*} \sum_w f(\beta)f(w|\beta)\log \frac{f(w|\beta)}{f(w|\bar{\mathscr{A}}^*)} \\
= & H_{\mathscr{A}^*} - H_{\min}
\end{aligned} \qquad(10)$$

We have now shown that for any fixed distributions $f(w|\mathcal{A})$ and $f(w|\bar{\mathcal{A}})$, applying rule (7) attains the minimum of the right-hand side of (8), and that replacing $f(w|\mathcal{A})$ and $f(w|\bar{\mathcal{A}})$ in that expression by $f(w|\mathcal{A}^*)$ and $f(w|\bar{\mathcal{A}}^*)$, respectively, makes that expression still smaller. As a result, $H_{\mathcal{A}} \geq H_{\mathcal{A}^*}$.

We have now the following (suboptimal) Chou's partitioning algorithm:

1. Choose atomic history subsets $\beta_1, \beta_2, \ldots, \beta_N$ of the set of histories $\mathcal{S}$ of the parent node.

2. On the basis of the chosen atomic subsets, partition $\mathcal{S}$ into some initially chosen complementary subsets $\mathcal{A}$ and $\bar{\mathcal{A}}$.

3. Compute

$$f(w|\mathcal{A}) = \frac{\sum_{\beta \in \mathcal{A}} f(w, \beta)}{\sum_{\beta \in \mathcal{A}} f(\beta)} \quad \text{and} \quad f(w|\bar{\mathcal{A}}) = \frac{\sum_{\beta \in \bar{\mathcal{A}}} f(w, \beta)}{\sum_{\beta \in \bar{\mathcal{A}}} f(\beta)} \tag{11}$$

4. For each $\beta \in \mathcal{S}$, if

$$\left[\sum_w f(w|\beta) \log \frac{f(w|\beta)}{f(w|\mathcal{A})} \right] \leq \left[\sum_w f(w|\beta) \log \frac{f(w|\beta)}{f(w|\bar{\mathcal{A}})} \right]$$

then place β into a new subset $\mathcal{A}^*$. Else place β into its complement $\bar{\mathcal{A}}^*$.

5. If $\mathcal{A}^* = \mathcal{A}$ then stop. Else set $\mathcal{A} = \mathcal{A}^*$ and $\bar{\mathcal{A}} = \bar{\mathcal{A}}^*$ and return to step 3.

10.8   Selection of the Initial Split of a Set $\mathcal{S}$ into Complementary Subsets

In general, the quality of the final partitions $\mathcal{A}$ and $\bar{\mathcal{A}}$ will depend on the partitions of $\mathcal{S}$ originally chosen in step 2 of Chou's partitioning algorithm. A good initial choice might be a set $\mathcal{A}$ composed of randomly chosen atoms $\beta \in \mathcal{S}$.

Note, however, that the initial partitioning of atoms can in principle be based on any pair of distributions $q_1(\ )$ and $q_2(\ )$. That is, at its first application, step 3 can be skipped and step 4 replaced by

4^{init}. For each $\beta \in \mathcal{S}$, if

$$\left[\sum_w f(w|\beta) \log \frac{f(w|\beta)}{q_1(w)} \right] \leq \left[\sum_w f(w|\beta) \log \frac{f(w|\beta)}{q_2(w)} \right] \tag{12}$$

then place β into a new subset $\mathcal{A}^*$. Else place β into its complement $\bar{\mathcal{A}}^*$.

Of course, if the two q-distributions are badly chosen then it can happen that $\mathscr{A}^* = \mathscr{S}$, that is, no split will take place.

10.9   The Two-ing Theorem

The Chou algorithm iteratively improves an original node split. No optimality claims can be made for it even under the restriction of an arbitrarily defined atom set. We simply choose starting distributions $q_1(\ )$ and $q_2(\ )$ and let the algorithm run. Under special circumstances, however, optimality can be guaranteed. The following theorem holds [1]:

THEOREM 10.1   THE TWO-ING THEOREM.   If the predicted variable has two values only[13] (i.e., 0 and 1), arrange the values of the predicting variables β_i so that $f(0|\beta_i) \geq f(0|\beta_{i+1})$. Then the optimal predicting set (relative to the entropy criterion) is

$$\mathscr{A} = \{\beta_j, j = 1, 2, \ldots, K\} \qquad \text{for some } K \tag{13}$$

Proof: We will use criterion (12) to prove the two-ing theorem.[14] To simplify the development, we will restate this criterion in terms of divergence functions:[15]

For each $\beta \in \mathscr{S}$, if

$$D(\mathbf{f}(\beta)\|\mathbf{q}_1) \leq D(\mathbf{f}(\beta)\|\mathbf{q}_2) \tag{14}$$

then place β into the subset $\mathscr{A}^*$. Else place β into its complement $\overline{\mathscr{A}^*}$.

In our proof we will use easy-to-prove convexity properties of the divergence function:[16] $D(\mathbf{f}\|\mathbf{q})$ is a convex function of either of its variables. That is:

1. $D(\lambda\mathbf{f}_1 + (1 - \lambda)\mathbf{f}_2\|\mathbf{q}) \leq \lambda D(\mathbf{f}_1\|\mathbf{q}) + (1 - \lambda)D(\mathbf{f}_2\|\mathbf{q})$

2. $D(\mathbf{f}\|\lambda\mathbf{q}_1 + (1 - \lambda)\mathbf{q}_2) \leq \lambda D(\mathbf{f}\|\mathbf{q}_1) + (1 - \lambda)D(\mathbf{f}\|\mathbf{q}_2)$

Let us assume, without loss of generality, that the binary distributions used in the splitting rule (14) are such that $q_1(0) < q_2(0)$.

---

13. Hence the name "two-ing".

14. The following use of the convexity properties of the divergence function was suggested by Sanjeev Khudanpur.

15. Definition (18) in section 7.4.

16. As usual, the proof depends on the inequality $\ln x \leq (x - 1)$.

First, any β such that $f(0|\beta) \leq q_1(0)$ will be placed in $\mathscr{A}^*$. In fact, because the distributions are binary, a value of $\lambda \in [0,1]$ will exist such that $\mathbf{q}_1 = \lambda \mathbf{f}(\beta) + (1 - \lambda)\mathbf{q}_2$. Applying the second of the above convexity properties, we get

$$D(\mathbf{f}(\beta)\|\mathbf{q}_1) \leq \lambda D(\mathbf{f}(\beta)\|\mathbf{f}(\beta)) + (1 - \lambda)D(\mathbf{f}(\beta)\|\mathbf{q}_2)$$

and since $D(\mathbf{f}(\beta)\|\mathbf{f}(\beta)) = 0$ then (14) holds and β is indeed placed into $\mathscr{A}^*$.

By exactly the same reasoning any β such that $f(0|\beta) \geq q_2(0)$ will be placed in $\overline{\mathscr{A}^*}$.

We are thus left with considering atoms β_1 and β_2 for which $q_1(0) < f(0|\beta_1) < f(0|\beta_2) < q_2(0)$. Our assertion is that the following situation is impossible:

$$D(\mathbf{f}(\beta_2)\|\mathbf{q}_1) \leq D(\mathbf{f}(\beta_2)\|\mathbf{q}_2) \quad \text{and} \quad D(\mathbf{f}(\beta_1)\|\mathbf{q}_2) \leq D(\mathbf{f}(\beta_1)\|\mathbf{q}_1)$$

In fact, we can find a value of λ for which $\mathbf{f}(\beta_1) = \lambda \mathbf{f}(\beta_2) + (1 - \lambda)\mathbf{q}_1$. Then by the first convexity property

$$D(\mathbf{f}(\beta_1)\|\mathbf{q}_1) \leq \lambda D(\mathbf{f}(\beta_2)\|\mathbf{q}_1) + (1 - \lambda)D(\mathbf{q}_1\|\mathbf{q}_1)$$

and therefore $D(\mathbf{f}(\beta_1)\|\mathbf{q}_1) < D(\mathbf{f}(\beta_2)\|\mathbf{q}_1)$.[17]

By exactly the same reasoning we can show that $D(\mathbf{f}(\beta_2)\|\mathbf{q}_2) < D(\mathbf{f}(\beta_1)\|\mathbf{q}_2)$. Therefore, if $D(\mathbf{f}(\beta_1)\|\mathbf{q}_2) \leq D(\mathbf{f}(\beta_1)\|\mathbf{q}_1)$ then $D(\mathbf{f}(\beta_2)\|\mathbf{q}_2) < D(\mathbf{f}(\beta_1)\|\mathbf{q}_1)$. Similarly, if $D(\mathbf{f}(\beta_2)\|\mathbf{q}_1) \leq D(\mathbf{f}(\beta_2)\|\mathbf{q}_2)$ then $D(\mathbf{f}(\beta_1)\|\mathbf{q}_1) < D(\mathbf{f}(\beta_1)\|\mathbf{q}_2)$.

We can therefore conclude that it is impossible for rule (14) to assign β_2 to set $\mathscr{A}^*$ and β_1 to $\overline{\mathscr{A}^*}$. It thus follows that $\mathscr{A}^* = \{\beta_j, j = 1, 2, \ldots, K\}$ for some K no matter what binary distributions $\mathbf{q}_1$ and $\mathbf{q}_2$ we start with. Consequently the best split (no matter how it was determined) will have the same property, because otherwise an iteration of the Chou algorithm could improve this split.
Q.E.D.

Note that we did not prove that the Chou algorithm will achieve the optimum split, only that it will have the form (13).

Section 10.13.2 will introduce practical application of the two-ing theorem after we discuss binary encoding of vocabulary words.

17. This is because $\lambda < 1$.

10.10   Practical Considerations of Chou's Method

10.10.1   Problem of 0s in the q-Distribution: The Gini Index

The initial q-distributions might be chosen by finding a pair of atoms β_1 and β_2 in $\mathscr{S}$ that maximally differ in distribution and then setting $q_1(w) = f(w|\beta_1)$ and $q_2(w) = f(w|\beta_2)$. The problem is that if, say, for some $v \in \mathscr{V}, q_1(v) = 0$ and $f(v|\beta) > 0$ then β will not be put into the $\mathscr{A}^*$ set no matter how closely the distributions $q_1(\ )$ and $f(\ |\beta)$ resemble each other otherwise.

The easiest remedy is to provide a "floor" to the q-distributions, that is, let

$$q_1(w) = \frac{f(w|\beta_1) + \varepsilon}{1 + |\mathscr{V}|\varepsilon}$$

Another, less ad hoc remedy is to replace entropy by another criterion of distribution purity: the *Gini index* [7] defined by

$$G(\mathbf{q}) = \sum_x q(x) \sum_{y \neq x} q(y) = 1 - \sum_x q(x)^2 \qquad (15)$$

Note that $G(\mathbf{q}) = 0$ if $q(x) = 1$ for some x (and for the same case $H(\mathbf{q}) = 0$) and $G(\mathbf{q}) \leq 1 - 1/|\mathscr{V}|$ (for $q(x) = 1/|\mathscr{V}|$ for all x). The Gini index has many other characteristics of entropy. For instance, it is a convex function of its argument. In fact,

$$
\begin{aligned}
&G(\theta\mathbf{p} + (1 - \theta)\mathbf{q}) - \theta G(\mathbf{p}) - (1 - \theta)G(\mathbf{q}) \\
&= \theta \sum p(x)^2 + (1 - \theta) \sum q(x)^2 - \sum [\theta p(x) + (1 - \theta)q(x)]^2 \\
&= \theta(1 - \theta) \sum [p(x)^2 + q(x)^2 - 2p(x)q(x)] \\
&\geq 0
\end{aligned}
\qquad (16)
$$

with equality if and only if $\mathbf{p} = \mathbf{q}$.

Using the Gini index, we want to split the set $\mathscr{S}$ into subsets $\mathscr{A}$ and $\bar{\mathscr{A}}$ so as to minimize (compare with (4))

$$G_{\mathscr{A}} \doteq f(\mathscr{A})G(\mathbf{f}(\ |\mathscr{A})) + f(\bar{\mathscr{A}})G(\mathbf{f}(\ |\bar{\mathscr{A}})) \qquad (17)$$

Following the reasoning of section 10.7 we proceed as follows. Minimizing (17) is equivalent to minimizing

$$G_{\mathscr{A}} - G_{\min} = \sum_{\beta \in \mathscr{A}} f(\beta)[G(\mathbf{f}(\ |\mathscr{A})) - G(\mathbf{f}(\ |\beta))]$$

$$+ \sum_{\beta \in \bar{\mathscr{A}}} f(\beta)[G(\mathbf{f}(\ |\bar{\mathscr{A}})) - G(\mathbf{f}(\ |\beta))]$$

$$= \sum_{\beta \in \mathscr{A}} f(\beta) \sum_{w} [f(w|\beta)^2 - f(w|\mathscr{A})^2]$$

$$+ \sum_{\beta \in \bar{\mathscr{A}}} f(\beta) \sum_{w} [f(w|\beta)^2 - f(w|\bar{\mathscr{A}})^2]$$

$$= \sum_{\beta \in \mathscr{A}} f(\beta) \sum_{w} [f(w|\beta) - f(w|\mathscr{A})]^2$$

$$+ 2 \sum_{\beta \in \mathscr{A}} f(\beta) \sum_{w} f(w|\beta)f(w|\mathscr{A})$$

$$- 2 \sum_{\beta \in \mathscr{A}} f(\beta) \sum_{w} f(w|\mathscr{A})^2 \tag{18}$$

$$+ \sum_{\beta \in \bar{\mathscr{A}}} f(\beta) \sum_{w} [f(w|\beta) - f(w|\bar{\mathscr{A}})]^2$$

$$+ 2 \sum_{\beta \in \bar{\mathscr{A}}} f(\beta) \sum_{w} f(w|\beta)f(w|\bar{\mathscr{A}})$$

$$- 2 \sum_{\beta \in \bar{\mathscr{A}}} f(\beta) \sum_{w} f(w|\bar{\mathscr{A}})^2$$

$$= \sum_{\beta \in \mathscr{A}} f(\beta) \sum_{w} [f(w|\beta) - f(w|\mathscr{A})]^2$$

$$+ \sum_{\beta \in \bar{\mathscr{A}}} f(\beta) \sum_{w} [f(w|\beta) - f(w|\bar{\mathscr{A}})]^2$$

where the last equality follows from the fact that

$$\sum_{\beta \in \mathscr{A}} f(\beta) \sum_{w} f(w|\beta)f(w|\mathscr{A}) = \sum_{w} f(w, \mathscr{A})f(w|\mathscr{A}) = f(\mathscr{A}) \sum_{w} f(w|\mathscr{A})^2$$

$$= \sum_{\beta \in \mathscr{A}} f(\beta) \sum_{w} f(w|\mathscr{A})^2$$

Since $\sum_{w}[f(w|\beta) - f(w|\mathscr{A})]^2$ is a nonnegative quantity, it follows from (18) that

$$G_{\mathscr{A}} - G_{\min} \geq \sum_{\beta \in \mathscr{A}^*} f(\beta) \sum_{w} [f(w|\beta) - f(w|\mathscr{A})]^2$$

$$+ \sum_{\beta \in \bar{\mathscr{A}}^*} f(\beta) \sum_{w} [f(w|\beta) - f(w|\bar{\mathscr{A}})]^2 \tag{19}$$

where $\beta \in \mathscr{A}^*$ provided

$$\sum_w [f(w|\beta) - f(w|\mathscr{A})]^2 \le \sum_w [f(w|\beta) - f(w|\bar{\mathscr{A}})]^2 \tag{20}$$

and otherwise $\beta \in \bar{\mathscr{A}}^*$. To conclude the argument that will lead to the partitioning algorithm for which we are aiming, note that for *any* pair of sets $\mathscr{A}$ and $\mathscr{A}^*$,

$$\sum_{\beta \in \mathscr{A}^*} f(\beta)[f(w|\beta) - f(w|\mathscr{A})]^2$$

$$= \sum_{\beta \in \mathscr{A}^*} f(\beta)[f(w|\beta) - f(w|\mathscr{A}^*) + f(w|\mathscr{A}^*) - f(w|\mathscr{A})]^2$$

$$= \sum_{\beta \in \mathscr{A}^*} f(\beta)[f(w|\beta) - f(w|\mathscr{A}^*)]^2$$

$$+ 2 \sum_{\beta \in \mathscr{A}^*} f(\beta)[f(w|\beta) - f(w|\mathscr{A}^*)][f(w|\mathscr{A}^*) - f(w|\mathscr{A})] \tag{21}$$

$$+ \sum_{\beta \in \mathscr{A}^*} f(\beta)[f(w|\mathscr{A}^*) - f(w|\mathscr{A})]^2$$

$$= \sum_{\beta \in \mathscr{A}^*} f(\beta)[f(w|\beta) - f(w|\mathscr{A}^*)]^2 + \sum_{\beta \in \mathscr{A}^*} f(\beta)[f(w|\mathscr{A}^*) - f(w|\mathscr{A})]^2$$

$$\ge \sum_{\beta \in \mathscr{A}^*} f(\beta)[f(w|\beta) - f(w|\mathscr{A}^*)]^2$$

In fact, the last equality follows because $\sum_{\beta \in \mathscr{A}^*} f(\beta)[f(w|\beta) - f(w|\mathscr{A}^*)] = f(\mathscr{A}^*)[f(w|\mathscr{A}^*) - f(w|\mathscr{A}^*)] = 0$. Using (21) we can conclude that if rule (20) defines $\mathscr{A}^*$ and $\bar{\mathscr{A}}^*$ then

$$G_A - G_{\min} \ge \sum_{\beta \in \mathscr{A}^*} f(\beta) \sum_w [f(w|\beta) - f(w|\mathscr{A})]^2$$

$$+ \sum_{\beta \in \bar{\mathscr{A}}^*} f(\beta) \sum_w [f(w|\beta) - f(w|\bar{\mathscr{A}})]^2$$

$$\ge \sum_{\beta \in \mathscr{A}^*} f(\beta) \sum_w [f(w|\beta) - f(w|\mathscr{A}^*)]^2$$

$$+ \sum_{\beta \in \bar{\mathscr{A}}^*} f(\beta) \sum_w [f(w|\beta) - f(w|\bar{\mathscr{A}}^*)]^2 = G_{\mathscr{A}^*} - G_{\min}$$

Hence we have Chou's partitioning algorithm based on the Gini index:

1. Choose atomic history subsets $\beta_1, \beta_2, \ldots, \beta_N$ of the set of histories $\mathscr{S}$ of the parent node.

2. On the basis of the above atomic subsets, partition $\mathscr{S}$ into some initial complementary subsets $\mathscr{A}$ and $\bar{\mathscr{A}}$.

3. Compute

$$f(w|\mathscr{A}) = \frac{\sum_{\beta \in \mathscr{A}} f(w,\beta)}{\sum_{\beta \in \mathscr{A}} f(\beta)} \quad \text{and} \quad f(w|\bar{\mathscr{A}}) = \frac{\sum_{\beta \in \bar{\mathscr{A}}} f(w,\beta)}{\sum_{\beta \in \bar{\mathscr{A}}} f(\beta)} \tag{22}$$

4. For each $\beta \in \mathscr{S}$, if

$$\sum_w [f(w|\beta) - f(w|\mathscr{A})]^2 \leq \sum_w [f(w|\beta) - f(w|\bar{\mathscr{A}})]^2$$

then place β into a new subset $\mathscr{A}^*$. Else place β into its complement $\bar{\mathscr{A}^*}$.

5. If $\mathscr{A}^* = \mathscr{A}$ then stop. Else set $\mathscr{A} = \mathscr{A}^*$ and $\bar{\mathscr{A}} = \bar{\mathscr{A}^*}$ and go to step 3.

In conclusion of this section let us stress that although the Gini index is useful in avoiding anomalies in question selection, the final aim of reducing the entropy of the test set remains. So the tree development algorithm of section 10.5 remains unchanged, that is, based on entropies. The Gini index is used to arrive at a proposed question via the Chou algorithm, but the value of this question is judged by the entropy it induces.

10.10.2 Equivalence Classification Induced by Decision Trees

Trigram language models imply a bigram equivalence classification of histories. If a particular bigram v', v'' never occurs in the training data, then the distribution $f(\ |v',v'')$ is not even defined. Equivalence classes induced by decision trees seemingly do not have this problem. They arise as a result of *yes/no* answers to a succession of questions of the type "Is $\mathbf{h} \in \mathscr{S}$?" Such questions will have answers for any history $\mathbf{h}$ whatever, regardless of whether $\mathbf{h}$ has been encountered in training data.

The problem is that step 4 of Chou's algorithm (section 10.7) assigns to set $\mathscr{A}$ only atoms β that have been seen in the training data, and assigns unseen atoms always to the complementary set $\bar{\mathscr{A}}$. If this situation is not properly handled, the probability $P(w|\Phi(\mathbf{h}))$ for unseen $\mathbf{h}$ may, when such a history is observed at run time, have very little relation to what actually happens.

One way to avoid this danger is either to distribute at random the unseen atoms to sets $\mathscr{A}$ and $\bar{\mathscr{A}}$ or to proceed as follows:

1. Define the complementary set $\bar{\mathscr{A}}$.

2. Let $\mathscr{A}'$ be the set of all atoms belonging to $\bar{\mathscr{A}}$ found in the development data.

3. Let $\mathscr{A}^-$ be the set of all atoms belonging to $\bar{\mathscr{A}}$ not *found* in the development data.[18]

4. Split the parent node into leaves $[\mathscr{A} \cup \mathscr{A}']$ and $\mathscr{A}^-$.

5. Split the node $[\mathscr{A} \cup \mathscr{A}']$ into leaves $\mathscr{A}$ and $\mathscr{A}'$.

6. To the leaf $\mathscr{A}^-$ assign the statistics of its parent node.

The above modification of Chou's tree-growing algorithm ensures that every history in the test set necessarily ends up at some definite leaf. What may be controversial is the statistics assignment to $\mathscr{A}^-$ in the last step above.

Indeed, that leaf, though it is not actually based on any data found in the development set, can be split further. If, for instance, the atoms used to split the parent node were defined by

$$\beta_v = \{\mathbf{h} = w_{-1}, w_{-2}, \ldots : w_{-m} = v, v \in \mathscr{V}, \mathbf{h} \in \mathscr{S}\}$$

then there is a subset $\mathscr{T} \subset \mathscr{V}$ such that $\mathscr{A}^- = \{\mathbf{h} = w_{-1}, w_{-2}, \ldots : w_{-m} \in \mathscr{T}, \mathbf{h} \in \mathscr{S}\}$ and questions can be asked about histories belonging to $\mathscr{A}^-$. It is unclear how to estimate the statistics of the resulting leaves from development data. Ad hoc solutions are possible: We can define new atoms

$$\beta_v = \{\mathbf{h} = w_{-1}, w_{-2}, \ldots : w_{-r} = v, v \in \mathscr{V}, \mathbf{h} \in \mathscr{S}\}, \qquad r \neq m$$

use them to split the parent history set $\mathscr{S}$, and assign the resulting histories and their statistics to the corresponding splits (ones that use the same questions) of $\mathscr{A}^-$.

In section 10.11 we will introduce a tree-building method that does not exhibit the disadvantage we have just attributed to the Chou algorithm.

10.10.3 Computationally Feasible Specification of Decision Tree Equivalence Classes

Given any history $\mathbf{h}$ we must be able to find the equivalence class the decision tree assigns to it, i.e., the leaf to which $\mathbf{h}$ belongs. In particular, let us

---

18. For instance, if (as in (3))

$$\beta_v = \{\mathbf{h} = w_{-1}, w_{-2}, \ldots : w_{-m} = v, v \in \mathscr{V}, \mathbf{h} \in \mathscr{S}\}$$

then β_v will belong to $\mathscr{A}^-$ if there is no $\mathbf{h} \in \mathscr{S}$ such that $w_{-m} = v$. This situation can easily happen.

consider the language modeling case where $\mathbf{h} = w_{-1}, w_{-2}, \ldots, w_{-m}, \ldots,$ and the question at the lth tree node is of the form "*Is* $w_{-m} \in \mathcal{T}_l$?"[19]

If the sets $\mathcal{T}_l$ are specified either by Chou's method or by two-ing, then the equivalence class specification may be quite complex, because the sets $\mathcal{T}_l$ for different values of l will be, in general, unrelated to each other. A straightforward solution[20] is to encode every word v of the vocabulary $\mathcal{V}$ into a bit string $b_1(v)b_2(v)\ldots b_M(v)$ using the following rule

$$b_l(v) = \begin{cases} 1 & \text{if } v \in \mathcal{T}_l \\ 0 & \text{otherwise} \end{cases}$$

Then the question at the lth node, instead of being "*Is* $w_{-m} \in \mathcal{T}_l$?" becomes "*Is* $b_l(w_{-m}) = 1$?" Of course, as before, with each node there must be associated the history position m about which the question will be asked.

10.11   Construction of Decision Trees Based on Word Encoding

As pointed out, the Chou method of tree construction runs into problems of unseen training data (section 10.2) and of question specification. In this section we will describe an attractive alternative specifically aimed at language modeling.[21] The idea is due to Mercer and is presented in Brown et al. [4].

Suppose the words of the given vocabulary $\mathcal{V}$ are classified into a full hierarchy of classes that can be specified by a binary tree whose leaves are the words themselves. Any node of the tree specifies a class that consists of the words at all the leaves of the subtree whose root is that node.

Figure 10.2 shows a possible tree for a vocabulary of size six. For instance, the node to which the double arrow points specifies the class consisting of the set of words $\{v_4, v_5, v_6\}$. As a result of such a hierarchical classification (one example of whose construction we will present in section 10.12), every word v_i can be encoded into a string of binary digits specifying which tree branches must be traversed to reach the leaf corresponding to v_i. Thus, for instance, v_5 is specified in figure 10.2 by the code word string 1011. (The code word when read from right to left specifies

---

19. See the discussion toward the end of section 10.10.2.

20. This solution was suggested by Mari Ostendorf.

21. The Chou method is totally universal.

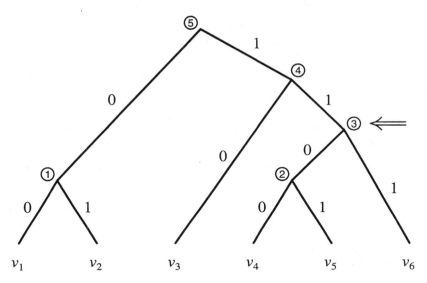

Figure 10.2
A word-encoding tree for a six-word vocabulary

the path from the tree root to the leaf; the reason for this apparently reversed direction will soon become clear.)

For the code words to be of equal length, we will conventionally pad them by leading 0s. Thus v_3 will correspond to 0001. Note that every suffix of any code word specifies a class to which the word belongs. We shall make use of this fact. As an example, in figure 10.2 the suffix 11 specifies the set $\{v_4, v_5, v_6\}$.

As a result of the word encoding, any history **h** consisting of N words can itself be represented by a string of $N \times L$ bits, where L is a single word's code length. The set of allowed questions (from which those actually determining the tree will be selected in the construction process) then concerns the identity of particular bits in the encoded history.

However, the identity of the ith bit of the kth code word (that is, the $(kL + i)^{th}$ bit counting from the right end of the word history bit string) by itself specifies no equivalence class. But if the suffix of length $i - 1$ of that code word is already known, then the ith bit refines further the class specified by the $i - 1$ previous bits. Therefore, questions about the ith code bit of a word will be allowed only if previous questions (along the path

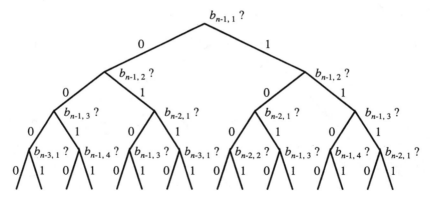

Figure 10.3
A decision tree language model based on binary encoding of words in the history

from the tree root) already ascertained the identity of the preceding $i - 1$ bits.

Figure 10.3 presents an example of an allowed tree-equivalence classification of a history. There we are predicting the nth word while b_{ji} ($j = n - 1, n - 2, \ldots, n - N; i = 1, \ldots, L$) denotes the ith bit of the jth word. Note that the twelfth leaf from the left is reached after the identity of the first three bits of the preceding and of the first bit of the penultimate word have been ascertained.

From the preceding two paragraphs it should be clear why we encoded from right to left: Laying the code words next to each other in natural text order will make the rightmost bits the most significant ones. Thus the first question will no doubt concern the first (rightmost) bit, although the quality of questions about bits $kL + 1$ for $k = 0, 1, \ldots, N - 1$ should also be investigated when the tree is being constructed. The quality of the language model obtained from a decision tree constructed by this method will obviously depend on whether the word encoding specifies word sets of interest.

10.12   A Hierarchical Classification of Vocabulary Words

In this section we describe a method, due to Mercer [8], of developing a hierarchical word classification (appropriate for language modeling) automatically from text data. The resulting tree will enable us to encode the word history into bit strings and will thus allow us to proceed with tree construction.

The basic idea is to develop a word classification Ψ that would result in a probability distribution $P(w_2|\Psi(w_1))$ whose average conditional entropy

$$H_\Psi = - \sum_{w_1,w_2} P(w_1, w_2)\log P(w_2|\Psi(w_1)) \tag{23}$$

would be minimal. Above, $P(w_1, w_2)$ denotes the joint probability of the two words w_1 and w_2 following each other in the text, and the sum is over all pairs w_1, w_2. Thus H_Ψ is a measure of the predictive power of the equivalence classification Ψ.

Since w_2 belongs to a unique equivalence class $\Psi(w_2)$,

$$P(w_2|\Psi(w_1)) = P(w_2, \Psi(w_2)|\Psi(w_1))$$

$$= P(w_2|\Psi(w_2), \Psi(w_1))P(\Psi(w_2)|\Psi(w_1))$$

Under the simplifying assumption that

$$P(w_2|\Psi(w_2), \Psi(w_1)) \cong P(w_2|\Psi(w_2)) \tag{24}$$

we then get

$$P(w_2|\Psi(w_1)) = P(w_2|\Psi(w_2))P(\Psi(w_2)|\Psi(w_1))$$

$$= \frac{P(w_2, \Psi(w_2))}{P(\Psi(w_2))} P(\Psi(w_2)|\Psi(w_1))$$

$$= P(w_2)\frac{P(\Psi(w_2)|\Psi(w_1))}{P(\Psi(w_2))}$$

where the last equality holds because $P(w_2, \Psi(w_2)) = P(w_2)$.

Therefore (23) becomes

$$H_\Psi = - \sum_{w_1,w_2} P(w_1, w_2)\left[\log P(w_2) + \log \frac{P(\Psi(w_2)|\Psi(w_1))}{P(\Psi(w_2))}\right]$$

$$= H(W_2) - \sum_{w_1,w_2} P(w_1, w_2)\log \frac{P(\Psi(w_2)|\Psi(w_1))}{P(\Psi(w_2))}$$

Since the entropy $H(W_2)$ is independent of any equivalence classification, then under the simplifying assumption (24), minimizing the conditional entropy H_Ψ is equivalent to maximizing the *mutual information function*

$$I_\Psi \doteq \sum_{w_1,w_2} P(w_1, w_2)\log \frac{P(\Psi(w_2)|\Psi(w_1))}{P(\Psi(w_2))} \tag{25}$$

Our method of hierarchical word classification will therefore strive to maximize I_Ψ. We proceed iteratively as follows:

1. Place all vocabulary words in their own separate class, thus obtaining K different classes, where $K = |\mathcal{V}|$ is the size of the vocabulary.
2. For each of the possible $K(K - 1)/2$ combinations of two classes into one resulting in a particular classification scheme of $K - 1$ classes, compute the corresponding mutual information I_Ψ. Choose that scheme whose I_Ψ was largest. Decrement K by 1.
3. If $K = 1$, stop. Else return to 2.

The above procedure obviously results in a hierarchical tree classification of the vocabulary of the type discussed in section 10.11. In the example of figure 10.2, the algorithm's first step resulted in the combination of words v_1 and v_2, forming the class of node 1. Then v_4 and v_5 were combined into the class of node 2. Next class 2 was combined with v_6 resulting in class 3. Then class 3 was combined with v_3, and class 4 was obtained. Finally, classes 1 and 4 were combined into the total class 5.

Figure 10.4 shows several actual classification tree segments resulting from the above procedure as reported in [8], which the reader should consult for more details. Section 10.13.1 offers a practical modification of the procedure making it computationally feasible.

10.13   More on Decision Trees Based on Word Encoding

10.13.1   Implementing Hierarchical Word Classification
With vocabulary sizes in the tens of thousands, it is practically impossible to try all the $\frac{1}{2}K(K - 1)$ class pairings as required in algorithm step 2. One way to speed things up is as follows [8].

Choose a size $M < K$, say $M \approx 1,000$. Then carry out the algorithm of section 10.12 restricted to the subset $\mathcal{V}_M$ of the vocabulary's M most frequent words (i.e., based on relative frequencies $f(v'|v)$; $v, v' \in \mathcal{V}_M$) until the number of classes is reduced to $\frac{M}{2}$. Then add the next most frequent $\frac{M}{2}$ words to the set, giving to all these words separate class designations (at which point there will be again M different classes), and continue the algorithm until the number of classes is once more reduced to $\frac{M}{2}$. Continue in this vein until all K vocabulary words have been "added" to the algorithm. Finally, complete the procedure to establish the root node of the classification tree.

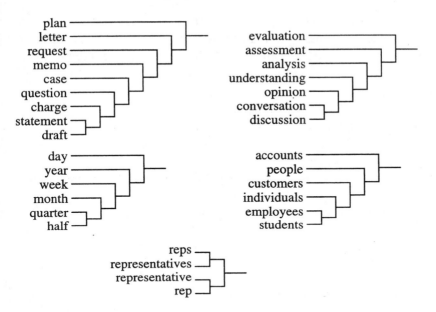

Figure 10.4
Sample subtrees from a 1,000-word mutual information word-encoding tree

10.13.2 Predicting Encoded Words One Bit at a Time

Let us observe that the word to be predicted can also be represented by its code word. Therefore, words can be predicted by predicting their individual code word bits, one after the other. Thus the language model could in principle be specified by L different trees,[22] each intended for the prediction of the next, i^{th} bit of the code word whose previous $i - 1$ bits have already been hypothesized.

Encoding of predicted words allows us to use the two-ing theorem in constructing a language model. In this case the words in the history would not be encoded, and use would be made of two-ing's optimality properties.[23]

---

22. L is the presumed length of the bit string encoding the words.

23. The ingenuity with which we select the basic atoms β on which the process is based limits its optimality. If the atoms have their "traditional" form $\beta_v = \{\mathbf{h} = w_{-1}, w_{-2}, \ldots : w_{-m} = v, v \in \mathcal{V}, \mathbf{h} \in \mathcal{S}\}$ then this limitation may be quite severe.

10.13.3   Treatment of Unseen Training Data

The problem of unseen training data that plagues the Chou method (section 10.10.2) does not arise when the history is encoded into bit strings and the questions concern the identity of bits in any particular position of the code words.

1. If all the data observed in the training text contain the same bit in a particular position, then that position will not be one used to split leaves, because a split would send all data in one direction only, and entropy would remain unchanged.

2. Though some strings having a particular bit in a particular position will not have been observed, it is reasonable to expect them to be correctly categorized. For instance, the bit-encoding (clustering) algorithm of section 10.12 assigns to words a particular bit prefix because they have been observed to be similar in their predictive power to other words with that same prefix.

10.13.4   Problems and Advantages of Word Encoding

The word-encoding method of set construction can alleviate the unseen-data problem encountered in the Chou algorithm, at least if the atoms are defined by

$$\beta_v = \{\mathbf{h} = w_{-1}, w_{-2}, \ldots : w_{-m} = v, v \in \mathscr{V}, \mathbf{h} \in \mathscr{S}\}$$

Using section 10.10.2's terminology, we need to place atoms β_v belonging to the set $\mathscr{A}^- = \{\mathbf{h} = w_{-1}, w_{-2}, \ldots : w_{-m} \in \mathscr{T}^-, \mathbf{h} \in \mathscr{S}\}$ either into $\mathscr{A} = \{\mathbf{h} = w_{-1}, w_{-2}, \ldots : w_{-m} \in \mathscr{T}, \mathbf{h} \in \mathscr{S}\}$ or into $\mathscr{A}' = \{\mathbf{h} = w_{-1}, w_{-2}, \ldots : w_{-m} \in \mathscr{T}', \mathbf{h} \in \mathscr{S}\}$.[24]

Now to every pair v, v^- corresponds the longest *suffix* $\mathbf{b}(v, v^-)$ that they share when encoded by the method of section 10.12. Furthermore, that suffix defines a set of words $\mathscr{V}(v, v^-)$ with the code suffix $\mathbf{b}(v, v^-)$. This is the set of words in the encoding tree dominated by the node corresponding to the suffix $\mathbf{b}(v, v^-)$.[25] With $v^- \in \mathscr{T}^-$ fixed, there is a word $v^* \in \mathscr{T} \cup \mathscr{T}'$ for which the set $\mathscr{V}(v^*, v^-)$ contains the smallest number of elements. One could well take the view that v^* is closest in "meaning" to

24. Note that $\mathscr{T} \cup \mathscr{T}' \cup \mathscr{T}^- = \mathscr{V}$ and that the three $\mathscr{T}$-sets have empty intersections.

25. For instance, in the encoding tree of figure 10.2, $\mathbf{b}(v_4, v_6) = 11$ and $\mathscr{V}(v_4, v_6) = \{v_4, v_5, v_6\}$.

v^- from among all of the words in the set $\mathcal{T} \cup \mathcal{T}'$. It is then logical to place β_{v^-} into $\mathcal{A}$ if $v^* \in \mathcal{T}$, and otherwise place it into $\mathcal{A}'$. This serves as the basis for deciding into which of two complementary sets one should place unseen atoms β_{v^-} when carrying out the Chou algorithm.

The encoding in section 10.12 was based on bigram statistics and thus seems very appropriate for defining sets of related-history words whose position immediately precedes that of the word being predicted. But the algorithm of section 10.11 uses the encoding as the basis of set selection for words in any history position, not just the last. It is an open problem to find a meaningful encoding that will underlie set selection for words in the former positions.

10.14   Final Remarks on the Decision Tree Method

In tree language modeling there are three distinct ways to overtrain:

1. Questions are assigned to nodes on the basis of insufficient data.
2. The decision tree continues to be refined beyond the natural limit the training data imposes.
3. The contents of the leaf sets $\mathcal{S}_{ij}$ are insufficient to allow a good estimate of the probabilities $P(w|\Phi_{ij})$.

In section 10.5 we introduced the method of cross-validation, in which we used a check set to prevent us from falling into the first two of the above traps. To alleviate the third problem, we must smooth. This means that the probabilities $P(w|\Phi_{ij})$ will be obtained as a linear combination of relative frequencies.

10.14.1   Smoothing

To show how to smooth, we must introduce some notation. Following the practice of section 10.11, each equivalence class can be specified by a binary code word designating the sequence of answers that lead to the corresponding leaf node. So, letting the answer *Yes* be designated by the bit 0 and the answer *No* by 1, the set Φ_{44} of figure 10.1 is specified by 100 and the set Φ_{45} by 101. If the length of the code word specifying Φ_{ij} is L, then Φ_{ij} is specified by a bit string $\mathbf{b}_1^L = b_1 b_2 \ldots b_L$, and any subsequence $\mathbf{b}_1^k, k = 1, \ldots, L$ designates a node along the tree path from the root to the leaf $\mathbf{b}_1^L$.

Now let $f(v|\mathbf{b}_1^k)$ be the relative frequency of the word v in the set $\mathcal{S}(\mathbf{b}_1^k)$ corresponding to the node $\mathbf{b}_1^k$. ($\mathcal{S}(\mathbf{b}_1^k)$ is then the union of all the sets corresponding to leaves that can be reached from $\mathbf{b}_1^k$.) The required smoothed probability is then

$$P_0(v|\Phi_{ij}) = P_0(v|\mathbf{b}_1^L) = \sum_{k=0}^{L} \lambda_k f(v|\mathbf{b}_1^k) \tag{26}$$

where the nonnegative weights satisfy $\sum_k \lambda_k = 1$. (We define $f(v|\mathbf{b}_1^0) \doteq f(v)$.) Obviously, λ_k should be estimated by the Baum algorithm carried out on check data when $f(v|\mathbf{b}_1^k)$ were calculated over separate development data. Furthermore, λ_k should be functions of the size $C(\mathbf{b}_1^k)$ of the sets $\mathscr{S}(\mathbf{b}_1^k)$.

One natural way to proceed is to estimate iteratively[26]

$$P_k(v|\mathbf{b}_1^L) = \gamma(C(\mathbf{b}_1^k))f(v|\mathbf{b}_1^k) + (1 - \gamma(C(\mathbf{b}_1^k)))P_{k+1}(v|\mathbf{b}_1^L) \tag{27}$$

for $k = L - 1, L - 2, \ldots, 0$ with the *boundary condition*

$$P_L(v|\mathbf{b}_1^L) \doteq f(v|b_1^L)$$

where $f(v|\mathbf{b}_1^0) \doteq f(v)$ denotes the relative frequency of v in the entire development set. Note that if the values $\gamma(c)$ are known for all counts c, then so are the values λ_k to be used in formula (26).

Formula (27) implies that the weights λ in (26) will be evaluated separately for each different leaf $\mathbf{b}_1^L$. This, of course, is not wise. Instead, one should tie all the coefficients $\gamma(C(\mathbf{b}_1^k))$ for the same tree depth k. As for all other ad hoc procedures, smoothing weights λ can be obtained in many other ways.

10.14.2 Fragmentation of Data

The problems pointed out in this section 10.14 as well as in section 10.10 stem from the fragmentation of data, a consequence of the splits that create the decision tree. Many attempts have been made to prevent or at least alleviate the problem. One is asking identical questions at different nodes. Another is the simultaneous development and combination of multiple parallel trees. This approach will be discussed in chapter 13. A fruitful and simple approach is as follows.

Suppose, based on the training data, we construct several trees. One way is to divide data into (possibly overlapping) K equal-sized segments and to construct an independent decision tree for each segment. These

---

26. This is the so-called *bottom-up* smoothing, which leads to better results than the more obvious top-down variety.

trees will have leaves to which are associated data sets $\mathscr{S}_{lj}$.[27] In a test situation, the language model will be presented with a history $\mathbf{h}$ and will be expected to supply the distribution $P(w|\mathbf{h})$. In the ith decision tree, this history will end up at some leaf $\Phi_i(\mathbf{h})$ corresponding to data set $\mathscr{S}_i(\mathbf{h})$. Let $f_i(w|\mathbf{h})$ be the relative frequency of the word w in the set $\mathscr{S}_i(\mathbf{h})$. Then the desired probability $P(w|\mathbf{h})$ can be based[28] on the "relative frequency"

$$f(w|\mathbf{h}) = \frac{1}{K} \sum_{i=1}^{k} f_i(w|\mathbf{h})$$

which is calculated on the fly.

10.15   Additional Reading

Broadly speaking, decision trees are used for learning how to classify data into prescribed types or for estimation of probability distributions. The first kind are referred to as *classification trees* and are typically created from training data whose class membership has been preannotated [9]. We have been dealing in this chapter with the second kind, *regression trees*. To estimate the required distributions we also partition data into classes (see section 10.3), but these are not prescribed a priori (rather they are discovered in the process), and the purpose of tree construction is not their identification.

In this chapter we concerned ourselves with splitting and stopping criteria. For the latter we advocated cross-validation [2] [10] [11],[29] but that is not the only way to proceed. A widely accepted alternative is to continue developing the tree (using a very mild stopping criterion) and then pruning back [1] [6] [13]. The prune-back criterion may again be based on cross-validation. Another possibility is to subject each proposed split to a statistical test of significance based on various assumptions about the data's distribution [14]. Finally, for some applications at least, it seems best to develop the tree fully and then linearly smooth the resulting distribution "up the tree," as was done in section 10.14.1 [4].

27. These sets form the basis of the relative frequency distributions $f(w|\Phi_{lj})$, as section 10.3 pointed out.

28. It is not equal to it, because smoothing is necessary, as always.

29. The article by Wolpert [12] describes a generalization of cross-validation.

A widely used method of data description is the famous *minimal description length* (MDL) principle [15] [16]. It is only natural to attempt to use it as a basis of tree organization, and in particular of the stopping criteria used [17] [18].

General problems of decision tree construction are considered by its pioneers [19] who also provide usable software [1] [20]. In a recent article, Breiman carries out a penetrating analysis of the properties of splitting criteria [21]. A heuristically satisfying method of splitting related to the two-ing theorem of section 10.10.9 can be found in reference [26]. The previously cited work by Martin [14] also analyzes splitting.

In many cases, the decision tree is being developed because its prospective users suspect the existence of fundamentally different modes of statistical behavior that they wish to segregate. In these cases the fast uncovering of alternate "pure" distributions is at issue [22].

Finally, several attempts to palliate the attendant fragmentation of data by simultaneous use of multiple tree have been published [23] [24] [25].

References

[1] L. Breiman, J.H. Friedman, R.A. Olshen, and C.J. Stone, *Classification and Regression Trees*, Wadsworth and Brooks, Pacific Grove, CA, 1984.

[2] M. Stone, "Cross-validatory choice and assessment of statistical predictions," *Journal of Royal Statistical Society*, ser. B, vol. 36, pp. 111–47, 1974.

[3] P.A. Chou, "Optimal partitioning for classification and regression trees," *IEEE Transactions on Pattern Analysis and Machine Intelligence*, vol. 13, no. 4, pp. 340–54, April 1991.

[4] P.F. Brown, S.A. Della Pietra, V.J. Della Pietra, R.L. Mercer, and P.S. Resnik, "Language modeling using decision trees," IBM Research Report Yorktown Heights, NY, 1991.

[5] L.R. Bahl, P.F. Brown, P.V. deSouza, and R.L. Mercer, "A tree-based language model for natural language speech recognition," *IEEE Transactions on Acoustics, Speech and Signal Processing*, vol. 37, pp. 1001–08, July 1989.

[6] P.A. Chou, T. Lookabaugh, and R.M. Gray, "Optimal pruning with applications to tree-structured source coding and modeling," *IEEE Transactions on Information Theory*, vol. 35, no. 2, pp. 299–315, March 1989.

[7] T.N. Bhargava and V.R.R. Uppuluri, "Sampling distribution of Gini's index of diversity," *Applied Mathematics and Computation*, vol. 3, pp. 1–24, 1977.

[8] P.F. Brown, V.J. Della Pietra, P.V. deSouza, J.C. Lai, and R.L. Mercer, "Class-based *n*-gram models of natural language," *Computational Linguistics*, vol. 18, no. 4, pp. 467–80, December 1992.

[9] P.A. Chou and R.M. Gray, "On decision trees for pattern recognition," *Proceedings of the IEEE Symposium on Information Theory*, p. 69, Ann Arbor, MI, October 1986.

[10] M. Stone, "Cross-validation: a review," *Statistics*, vol. 9, no. 1, pp. 127–39, 1978.

[11] M. Stone, "Cross-validation and multinomial prediction," *Biometrika*, vol. 61, no. 3, pp. 509–15, 1974.

[12] D.H. Wolpert, "Stacked generalization," *Neural Networks*, vol. 5, pp. 241–59, 1992.

[13] J. Mingers, "An empirical comparison of pruning methods for decision tree induction," *Machine Learning*, vol. 4, no. 2, pp. 227–43, 1990.

[14] J.K. Martin, "An exact probability metric for decision tree splitting and stopping," *Artificial Intelligence and Statistics*, vol. 5, pp. 379–85, 1995.

[15] J. Rissanen, *Stochastic Complexity in Statistical Inquiry*, World Scientific Publishing Co., Teaneck, NJ, 1989.

[16] J. Rissanen, "Stochastic Complexity and Modeling," *Annals of Statistics*, vol. 14, pp. 1080–1100, September 1986.

[17] J.R. Quinlan and R. Rivest, "Inferring decision trees using the minimum description length principle," *Information and Computation*, vol. 80, no. 3, pp. 227–48, 1989.

[18] M. Wax, "Construction of tree structured classifiers by the MDL principle," *Proceedings of the IEEE International Conference on Acoustics, Speech and Signal Processing*, pp. 2157–60, Albuquerque, NM, April 1990.

[19] J.R. Quinlan, "Probabilistic decision trees," in *Machine Learning: An Artificial Intelligence Approach*, vol. III, R.S. Michalski and Y. Kodratoff, eds., pp. 140–52, Morgan Kaufmann, San Mateo, CA, 1990.

[20] J.R. Quinlan, "Induction of decision trees," *Machine Learning*, vol. 1, pp. 81–106, 1986.

[21] L. Breiman, "Technical note: some properties of splitting criteria," *Machine Learning*, vol. 24, pp. 41–47, 1996.

[22] D. Burshtein, V. Della Pietra, D. Kanevsky, and A. Nadas, "Minimum impurity partitions," *The Annals of Statistics*, vol. 20, no. 3, pp. 1637–46, 1992.

[23] L. Breiman, *Stacked Regressions*, Technical Report no. 421, University of California at Berkeley, September 1994.

[24] D. Heath, S. Kasif, and S. Salzberg, "Committees of decision trees," in *Cognitive Technology: In Search of a Humane Interface*, B. Gorayska and J. Mey, eds., Elsevier Science B.V., Amsterdam, 1996.

[25] L. Breiman, *Bagging Predictors*, Technical Report no. 367, University of California at Berkeley, August 1992.

[26] A. Nadas, D. Nahamoo, M.A. Picheny, and J. Powell, "An iterative 'flip-flop' approximation of the most informative split in the construction of decision trees," *Proceedings of the IEEE International Conference on Acoustics, Speech and Signal Processing*, vol. I, pp. 565–68, Toronto, May 1991.

Chapter 11

Phonetics from Orthography: Spelling-to-Base Form Mappings

11.1   Overview of Base Form Generation from Spelling

In this and the following chapters we will give two examples of using decision trees in constructing acoustic models. Many other applications are possible; the employment of decision trees is widespread. We selected the current instances not just for their importance, but also because they illustrate considerable ingenuity leading to self-organization from data.

Most speech recognizers need a phonetic base form for every word in the vocabulary:

1. for carrying out the fast match (chapter 6) concerning words in the basic vocabulary the system provides
2. for the detailed match, if the word HMM depends on the base form (chapters 3 and 12)
3. For customizing the vocabulary: to allow users to add words they choose by simply spelling them and providing a single instance of their pronunciation.

The problem then arises of how to generate the required base forms. A solution for it is particularly urgent in English where word spelling does not determine the pronunciation or even reliably indicate it, as it would in most other European languages.

In this chapter we will develop statistical spelling–to–base form mappings from data consisting of an initial (sufficiently large) phonetic dictionary that provides the correspondence between a word's spelling and its base form. To develop the mappings, we will use decision trees.

The basic idea of the approach is very simple [1].

1. For each word spelling–word base form pair in the phonetic dictionary, we will obtain an alignment of letters and phone strings "caused" by the

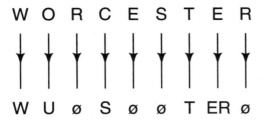

Figure 11.1
Alignment of the letters of the spelled word *Worcester* with the phones of its base form

letters. As a result, in particular situations a letter y might be aligned with the null phone ϕ having no acoustic realization, or with a string of phones $\varphi_1\varphi_2\ldots\varphi_k$, $k \geq 1$. Figure 11.1 illustrates this for the word *Worcester*.

2. The mapping $y \rightarrow \varphi_1\varphi_2\ldots\varphi_k$ takes place in an environment Ψ, $\mathbf{X}\_\_\mathbf{Z}$, where Ψ denotes the phone string preceding $\varphi_1\varphi_2\ldots\varphi_k$, $\mathbf{X}$ denotes the string of letters preceding y, and $\mathbf{Z}$ denotes the string of letters following y.

3. Using decision trees, we will develop an equivalence classification $\Theta_y(\Psi, \mathbf{X}\_\_\mathbf{Z})$ of environments of the letter y and estimate the probability of the mapping $y \rightarrow \varphi_1\varphi_2\ldots\varphi_k$ in the environment Ψ, $\mathbf{X}\_\_\mathbf{Z}$,

$$P(\varphi_1\varphi_2\ldots\varphi_k|y, \Theta_y(\Psi, \mathbf{XZ})) \tag{1}$$

We thus assign probabilities to the different phonetic realizations $\varphi_1\varphi_2\ldots\varphi_k$ of the letter y in the environment class $\Theta_y(\Psi, \mathbf{X}\_\_\mathbf{Z})$.

4. The *actual* construction of the base form $\hat{\Phi}(v) = \hat{\varphi}_1\hat{\varphi}_2\ldots\hat{\varphi}_h$ for the word v will depend on its sample pronunciation (acoustic string) $\mathbf{A} = a_1, a_2, \ldots, a_l$.[1] The orthographic spelling of v is denoted by $\mathbf{y}(v) = y_1, y_2, \ldots, y_m$. The base form will be computed by the formula[2]

$$\hat{\Phi}(v) = \arg\max_{\Phi} P(\mathbf{A}|\Phi)P(\Phi|\mathbf{y}(v)) \tag{2}$$

1. In a language like English it is not possible to derive the base form without a sample of its pronunciation. For instance, without an acoustic sample it is not possible to decide between the LÍD and LED pronunciations of the homographic word *lead*.

2. The base form $\hat{\Phi} = \hat{\varphi}_1\hat{\varphi}_2\cdots\hat{\varphi}_h$ will be a concatenation of the phone substrings $\varphi_1^i\varphi_2^i\cdots\varphi_k^i$ pertaining to the successive letters y_i in the word's spelling.

where $P(\mathbf{\Phi}|\mathbf{y}(v))$ is based on the probabilities (1) and $P(\mathbf{A}|\mathbf{\Phi})$ is obtained from the HMM of the word v which itself is a concatenation of basic HMMs determined by the phones of the string $\mathbf{\Phi}$.

11.2 Generating Alignment Data

We will align phone strings $\mathbf{\Phi}$ with letter strings $\mathbf{y}$ in two steps. In the first step we will pretend that $\mathbf{y}$ is generated from $\mathbf{\Phi}$ ($\mathbf{\Phi} \rightarrow \mathbf{y}$) and find the most likely alignment between the two, and in the second step we will construct an alignment with a reversed cause and effect, $\mathbf{y} \rightarrow \mathbf{\Phi}$.[3]

The first step will be based on an HMM whose observables are strings $y_1, y_2, \ldots, y_m$. The HMM itself is the natural concatenation of basic HMMs that correspond to the individual phones φ_i, $i = 1, 2, \ldots, n$ making up the base form $\mathbf{\Phi}$.[4] The basic HMM has the structure of figure 11.2, and it generates strings with probability

$$P(y_1, y_2, \ldots, y_m|\varphi) = \begin{cases} R(m|\varphi) \prod_{i=1}^{m} r(y_i|\varphi) & \text{for } 1 \le m \le 5 \\ R(0|\varphi) & \text{for } m = 0 \end{cases} \tag{3}$$

In figure 11.2 we have marked the transition probabilities $R(m|\varphi)$. All solid transitions are assumed to have the same output distribution $r(y|\varphi)$.

Estimates $R(m|\varphi)$ and $r(y|\varphi)$ are obtained by applying the Baum algorithm (section 2.7) to the data constituted by the initial phonetic dictionary. That is,

1. Initialize appropriately the parameters of the various building block HMMs (figure 11.2 indicates its structure).

2. For all $\mathbf{\Phi}(v)$, $\mathbf{y}(v)$ combinations found in the lexicon, construct the letter-generating word HMM by concatenating the appropriate building blocks specified by $\mathbf{\Phi}(v)$. This HMM is presumed to have generated the "data" $\mathbf{y}(v)$.

3. We do not carry out the desired alignment $\mathbf{y} \rightarrow \mathbf{\Phi}$ directly because in English it is much more likely that a phone φ is realized by a string of letters $y_1, y_2, \ldots, y_m$, $m \ge 2$, than that a letter is realized by a string of phones. Moreover, even though we provide for it, a phone φ would seldom map into no letter at all.

4. Triphone (section 3.4 and chapter 12) or any other acoustic model building blocks specified by $\mathbf{\Phi}$ can be used, but phone building blocks seem to do the job.

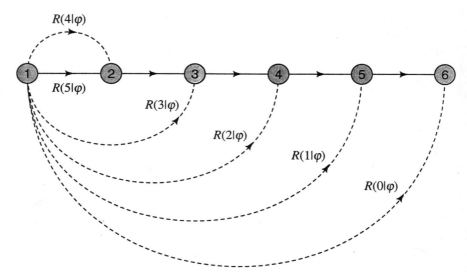

Figure 11.2
Basic letter generating HMM that corresponds to the phone φ

3. In the obvious manner, apply the Baum algorithm to the entire data $\mathbf{y}(v)$, $v \in \mathscr{V}$.[5] Thus estimate the building block parameters $R(m|\varphi)$ and $r(y|\varphi)$.

Once training is complete, the Viterbi algorithm is applied individually to each $\mathbf{\Phi}(v)$, $\mathbf{y}(v)$ combination. This results in the most likely alignment of phones with letters in the spelling of v (according to the estimated models (3)). Figure 11.3 illustrates this for the word *Worcester*. Note, for instance, that the phone s is conceptualized there as having given rise to the letter string CES.

We next create the opposite alignment, in which letters give rise to phones, by a simple rule:

$$\text{if } \varphi \to y_1, y_2, \dots, y_m \quad \text{then } y_1 \to \varphi \text{ and } y_i \to \phi, \quad 1 < i \le m$$
$$\text{if } \varphi \to \phi \qquad\qquad\qquad \text{then } \phi \to \varphi \tag{4}$$

Figure 11.1 shows the result of applying rule (4) to figure 11.3.

5. A pass through all the data constitutes a single iteration.

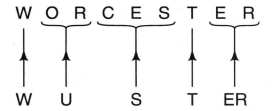

Figure 11.3
Automatically obtained alignment of the word *Worcester* with the phones of its base form

11.3 Decision Tree Classification of Phonetic Environments

The alignments of the preceding section have provided us with data that constitute the basis for deriving the equivalence classifier $\Theta_y(\Psi, X\_\_Z)$ for each different letter y, including the null letter ϕ.

From these alignments we get (by rule (4)) for each y a list of mappings

$$\Psi, X\_y\_Z \to \varphi \tag{5}$$

where φ is the phone that y produced in the environment $\Psi, X\_\_Z$.

Using this data we will derive, for each letter y, a decision tree based on questions (a) about set memberships of phones found in particular positions of the (already realized) prefix string Ψ; and (b) about set memberships of letters found in particular positions of the strings X and Z that provide the orthographic context of the letter y. These strings are specified in the order of adjacency to y, so in the example of figure 11.1, if $y = s$, then $\varphi = \phi$, $\Psi = \text{suw\#}$, $X = ecroW\#$, and $Z = ter\#$, where we introduced the end-of-word markers $\#$.

The criterion of question selection is the usual one of entropy, and we will divide the data into development and check portions to be able to invoke the stopping rule of section 10.5. The list of candidate sets that underlie the question selection process[6] includes the natural phonetic ones: individual phones (letters), vowels, consonants, fricatives, stops, labials, etc.[7]

---

6. Recall the beginning of section 10.6.

7. Of course, the sets could also be constructed via the Chou algorithm. This risks running into the problems discussed in section 10.14. The natural phonetic classes defined for phones can, after a minimum of thought, also be applied to letters.

To simplify our work, we can truncate the strings $\boldsymbol{\Psi}$, $\mathbf{X}$, and $\mathbf{Z}$ to the four phones or letters most adjacent to the produced phone φ or the producing letter y. We can then denote the resulting symbol string $\boldsymbol{\Psi}$, $\mathbf{X\_\_Z}$ by the history designator $\mathbf{h}$ and thus reduce the problem to that encountered in chapter 10, where we discussed tree language models. For added clarity, we will restate the procedure.

For the purposes of tree construction, let us define the probabilities

$$P(\varphi|\Theta_{ij}) \doteq f(\varphi|S_{ij})$$

where the subscripts i and j refer to the jth branch on the ith level of the decision tree, S_{ij} and Θ_{ij} refer to the corresponding data set and equivalence class, respectively (section 10.3), and f denotes the relative frequency of the phone φ in the corresponding data set,

$$f(\varphi|S_{ij}) \doteq \frac{C(\varphi|S_{ij})}{C(S_{ij})}$$

Above, $C(S_{ij})$ denotes the number of mappings from the list (5) that the decision tree directed to the set S_{ij}. Finally, let

$$f(S_{ij}) \doteq \frac{C(S_{ij})}{n}$$

where n is the number of entries in the mapping data set (5) pertaining to the letter y. The average conditional entropy of any particular development tree is then given by

$$H = -\sum_{i,j} \sum_{\varphi} f(S_{ij}) f(\varphi|\Theta_{ij}) \log P(\varphi|\Theta_{ij})$$

and that of the check tree by the cross entropy

$$H^c = -\sum_{i,j} \sum_{\varphi} g(S_{ij}) g(\varphi|\Theta_{ij}) \log P(\varphi|\Theta_{ij})$$

where the sum is over all the leaves i, j of the current tree (of course, f and g denote the relative frequencies corresponding to the development and check trees, respectively).

The entropy H given above is the proposed goodness criterion. Before stating the tree construction algorithm, we point out again that questions Q refer to set membership of symbols in particular positions of $\mathbf{h}$.

1. From among all possible questions, find that question Q_{10} that will classify all histories $\mathbf{h}$ into one of two equivalence classes Θ_{20} and Θ_{21} such

that the resulting value of H computed on the development set is minimal. Call this value H_d. Using question Q_{10} evaluate H^c over the check set.

2. Keeping Q_{10} fixed, find the best tentative question Q_{20} that will help split histories **h** into three equivalence classes Θ_{30}, Θ_{31}, and Θ_{21} such that the resulting value of H computed on the development set is minimal. Call this value H_{20}.

3. Keeping Q_{10} fixed, find the best tentative question Q_{21} that will help split histories **h** into three equivalence classes Θ_{20}, Θ_{32}, and Θ_{33} such that the resulting value of H computed on the development set is minimal. Call this value H_{21}.

4. Of Q_{20} or Q_{21}, eliminate the one that results in a larger value of H. Call the remaining question Q_{2j}. If $H_d - H_{2j} < \varepsilon$ then stop and reject Q_{2j} as well. Otherwise evaluate H_{2j}^c over the check set. If $H^c - H_{2j}^c < \delta$ then stop. Else make question Q_{2j} permanent.

5. The pattern is now set. Assuming that Q_{21} was chosen, we will next choose tentative questions Q_{20}, Q_{32}, and Q_{33}, and make permanent the one that lowers that entropy of the development set by at least ε and, simultaneously, that of the check set by at least δ. If both these improvements do not take place, no additional question is accepted and the tree construction process terminates.

6. And so forth, continuing the pattern set.

Once the tree is constructed,[8] the sets S_{ij} corresponding to its final leaves are recalculated on the basis of all of the training data (i.e., of the combined development and check sets).

It should be noted that what we derive are probabilities

$$P(\varphi|y, \Theta_y(\mathbf{h})) = P(\varphi|y, \Theta_y(\mathbf{\Psi}, \mathbf{X\_\_Z}))$$

for all letters y including the empty letter ϕ. Therefore, we can calculate the probabilities

$$P(\varphi_1\varphi_2 \ldots \varphi_k|y, \Theta_y(\mathbf{h})) = P(\varphi_1|y, \Theta_y(\mathbf{\Psi}, \mathbf{X\_\_Z}))$$

$$\times \prod_{i=2}^{k} P(\varphi_i|y, \Theta_y(\mathbf{\Psi}_i', \mathbf{X\_\_Z})) \tag{6}$$

---

8. As was pointed out in section 10.5, the stopping rule invoked here is probably unnecessarily conservative. We have indicated there how it can be loosened. We also mentioned there that many alternative stopping rules based strictly on development data can also be used.

where we define $\mathbf{\Psi}'_i \doteq \varphi_{i-1} \ldots \varphi_1 \mathbf{\Psi}$.[9] Determination of probabilities (6) was the aim of this section.

11.4 Finding the Base Forms

As stated in the section 11.1, the desired base form for word v is given by the formula

$$\hat{\mathbf{\Phi}}(v) = \arg \max_{\mathbf{\Phi}} P(\mathbf{A}|\mathbf{\Phi})P(\mathbf{\Phi}|\mathbf{y}(v)) \qquad (7)$$

where $\mathbf{A}$ is the acoustic string corresponding to the sample utterance of v. Thus the search for the solution is formally identical to the speech recognition search for words the user speaks. The only difference is that in the latter case, instead of $P(\mathbf{\Phi}|\mathbf{y}(v))$ we had $P(\mathbf{W})$, and instead of $P(\mathbf{A}|\mathbf{\Phi})$ we had $P(\mathbf{A}|\mathbf{W})$. Therefore, to find $\hat{\mathbf{\Phi}}$ we employ either the stack or the Viterbi algorithms (chapters 6 and 5). The branches of the associated hypothesis tree (figure 6.1) are associated with the individual phones φ.

The remaining question is how to compute the language model probabilities $P(\mathbf{\Phi}|\mathbf{y}) = P(\varphi_1 \varphi_2 \ldots \varphi_i|\mathbf{y})$. This is not completely trivial, because in principle the individual letters y_i making up the word spelling $\mathbf{y}$ can generate a string of phones of any length (see (6)).

The required computation can best be carried out on a trellis whose i^{th} stage is associated with the individual phone φ_i. Figure 11.4 is an example for the word $y_1 y_2 y_3$ producing the string $\varphi_1 \varphi_2$. The trellis states are indicated in the form $\mathbf{\Psi}|\mathbf{X}|y|\mathbf{Z}$, where both $\mathbf{X}$ and $\mathbf{\Psi}$ are, to make reading easier, stated in the natural sequence. The required probabilities $P(\varphi_1 \varphi_2 \ldots \varphi_i|\mathbf{y})$ are then given by the sum of probabilities associated with the states in the trellis's i^{th} column.

11.5 Additional Reading

The original attempt to derive phonetic base forms from orthographic spelling by means of decision trees is due to Lucassen and Mercer [2]. This work is the basis for the approach of reference [1] .

The problem is of great importance to speech synthesis (text-to-speech systems) [3] and has therefore been studied for some time [4]. A recent approach based on rules and data is due to Meng et al. [5]. The most up-

---

9. Note that in (6) we predict phones from left to right, so by the time it is the turn of φ_i, the string $\varphi_1 \varphi_2 \ldots \varphi_{i-1}$ of preceding phones provides the context.

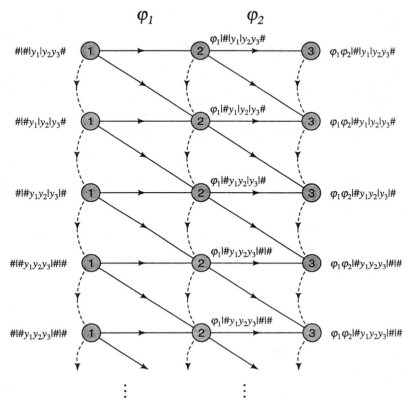

Figure 11.4
Trellis used in computing the probability of the letter string $y_1y_2y_3$ generating the phone string $\varphi_1\varphi_2$ (trellis states are indicated in the form $\Psi|\mathbf{X}|y|\mathbf{Z}$, where Ψ denotes the already generated phone string and $\mathbf{X}$ and $\mathbf{Y}$ are left and right letter string contexts of the letter y)

to-date results can be found in the work of Luk and Damper [6] which uses a finite state stochastic grammar as a model whose parameters are inferred from a pronouncing dictionary. Their article contains an extensive bibliography.

References

[1] L.R. Bahl, S. Das, P.V. deSouza, M. Epstein, R.L. Mercer, B. Merialdo, M.A. Picheny, and J. Powell, "Automatic phonetic baseform determination," *Proceedings of the 1991 International Conference on Acoustics, Speech, and Signal Processing*, pp. 173–76, Toronto, May 1991.

[2] J.M. Lucassen and R.L. Mercer, "An information theoretic approach to the automatic determination of phonemic baseforms," *Proceedings of the 1984 International Conference on Acoustics, Speech, and Signal Processing*, vol. III, pp. 42.5.1–42.5.4, San Diego, CA, March 1984.

[3] C. Sorin, "Towards high-quality multilingual text-to-speech," in *Progress and Prospects of Speech Research and Technology*, H. Niemann, R. de Mori, and G. Hanrieder, eds., pp. 53–62, Infix Publishing Co., Sankt Augustin, Germany, 1994.

[4] C. Coker, "A dictionary-intensive letter-to-sound program," *Journal of the Acoustic Society of America*, vol. 78, Supp. 1, S7, 1985.

[5] H.M. Meng, H. Seneff, and V.W. Zue, "Phonological parsing for reversible letter-to-sound/sound-to-letter generation," *Proceedings of the 1994 International Conference on Acoustics, Speech, and Signal Processing*, vol. II, pp. 1–4, Adelaide, South Australia, April 1994.

[6] R.W.P. Luk and R.I. Damper, "Stochastic phonographic transduction for English," *Computer Speech and Language*, vol. 10, no. 2, pp. 133–53, April 1996.

Chapter 12

Triphones and Allophones

12.1 Introduction

Speech production obviously cannot be accurately modeled by a concatenation of elementary HMMs corresponding to individual phones of a word base form. The realization of phones depends on their context. This is especially true for initial and final segments of words in continuous speech that are influenced by the phenomenon known as *coarticulation*. Indeed, the **t** phone is aspirated in a word like *top*, but lacks aspiration in *pot*. We also noticed when discussing the speech waveform for *Bishop moves to king knight five* (figure 1.1), that although the two **i** phones seem to sound the same, the appearance of the waveform of the first **i** differs markedly from that of the second.

Phoneticians speak of the desirability of a more refined *allophonic* alphabet [1] that would contain such elements as a *nasalized* **r** or a *retroflexed* **l**. But how many allophones are there and which ones? And in which contexts are they used?

In this chapter we will discuss several methods that will yield HMM building blocks that take into account phonetic context. We will employ decision trees to achieve the equivalence classification the process requires.

We would like to be able to specify a word by its ordinary phonetic base form and then transform the base form's phone elements into their appropriate allophones according to the context in which the phones find themselves. To each allophone would correspond an elementary HMM, and a concatenation of these building blocks, specified by the word base form, would then result in the word HMM.

Note that achieving the above goal would have three important consequences:

1. Knowing the phonetic base form of a new word would allow us to model the word accurately without collecting speech data for it, as is necessary for fenonic baseforms (sections 3.6 and following).
2. The spelling-to-sound rules introduced in chapter 11 would remain an adequate means for users to personalize the vocabulary.
3. The addition of an end-of-word phone marker # to the phonetic alphabet could become a means for handling coarticulation in continuous speech. As an example, the phrase *press show* would result in the phone string # PRES # ŠOU #, and if an allophonic rule converted s in the context E_ _ # š into the null allophone, the phrase would get the pronunciation PREŠOU.

In section 12.2 we will deal with triphone building blocks, which we discussed rather superficially in section 3.4. Sections 12.3 through 12.6 will introduce a sophisticated use of decision trees leading to general contextual building blocks. At the end of that discussion, in section 12.7, we will return to triphones as they are treated in current state-of-the-art multispeaker recognition systems [6].

12.2   Triphones

Most current speech recognizers base their modeling on the triphone concept [2] [3] [4]: The allophone of φ embedded in the context $\ldots \varphi_1 - - \varphi_2 \ldots$ is simply specified by the triphone $\varphi_1 \varphi \varphi_2$. But if this solution were carried out literally, the resulting allophone alphabet would be too large to be useful. A phonetic alphabet of size $M \approx 60$ would result in $M^3 \approx 216,000$ allophones. The corresponding HMM building blocks would take up too much storage and, of course, there would never be enough data to train them.

So the triphone method may have to rely on an equivalence classification $\Phi(\varphi_1 \varphi \varphi_2)$. The following sections will consider how to achieve such a classification, but here is a simple and quite effective approach:

1. Having decided on the total vocabulary $\mathscr{V}$ and having established the corresponding base form lexicon, determine the total possible inventory $\mathscr{T}$ of triphones.
2. Determine the structure of triphone HMMs. Number the transitions and denote by $t_i(\sigma)$ the ith transition of the HMM for the triphone σ. Thus $t_i(\sigma)$ and $t_i(\sigma')$ are in *structural* correspondence.

3. Get training speech $\mathbf{A}$ and its transcription $\mathbf{W}$. Construct the composite HMM for $\mathbf{W}$ using the triphone HMMs as building blocks. Given any value of i and φ, tie the transitions $t_i(\varphi_1\varphi\varphi_2)$ and $t_i(\varphi_1'\varphi\varphi_2')$ for all $\varphi_1, \varphi_2, \varphi_1', \varphi_2'$ and train.

4. Untie the transitions and continue training (i.e., use the HMM tied parameter values obtained in step 3 as initial probabilities for this second round of training). For each transition $t_i(\sigma)$ observe the number of times, $c(t_i(\sigma))$, it has been visited. (In general, this will not be an integer—we are talking here about accumulator contents in the Baum algorithm.)

5. Establish a threshold τ. Fixing each pair i and φ, tie together all transitions $t_i(\varphi_1\varphi\varphi_2)$ for those values of φ_1 and φ_2 for which $c(t_i(\varphi_1\varphi\varphi_2)) < \tau$. If the sum of the counts of transitions thus tied does not exceed the threshold τ, increase the set of tied transitions by the next-lowest-count transition $t_i(\varphi_1'\varphi\varphi_2')$ for some pair φ_1', φ_2' (by definition, because it was so far excluded from being tied, the count for this transition exceeds τ).

6. Retrain the new composite HMM. The resulting building block HMMs will serve as the recognizer's triphone models.[1]

It is perhaps worth remarking that most practical recognizers make use of additional building blocks. Thus in addition to the triphones, specific word models for short, frequent words such as *the, in*, etc., are constructed.

The reader may also have noticed that triphones, as we have treated them, would not handle the coarticulation problem illustrated by the *press show* phrase.[2] One way out would be to enlarge the phone alphabet by adding initial phones φ^i and eliminating the word boundary marker. The corresponding phone string would then be PRESŠ^{i}OU and the triphone model for ESŠ^i when trained might have silence-like characteristics.[3]

---

1. Though we don't mention it explicitly, smoothing of distributions will always be necessary and must be carried out in addition to the above tying process. Simple smoothing would use deleted interpolation (see sections 4.4 and 4.6) to combine the triphones of step 6 with the monophones of step 3.

2. It is true that in principle the phone string #PRES#ŠOU# could lead to successive building blocks ..., ES#, S#Š, #ŠO, ... and that the triphones ES# and S#Š could both have null realizations, whereas in the case of *press conference* the corresponding triphones ..., ES#, S#K, #KA, ... could have the realizations ES# $\rightarrow \phi$, S#K $\rightarrow s$, #KA $\rightarrow k$. But the training procedure of this section could not achieve such a result.

3. At the same time, SŠ^{i}O would have the characteristics of the Š^i PHONE.

12.3 The General Method

As pointed out, our problem can be regarded as deriving the appropriate equivalence classification of the context of phones. This will be done here on the basis of actual *speech data*, and the vehicle will again be decision trees [5].

Roughly speaking, for each phone of the basic phonetic alphabet we would like to:

1. Collect a great deal of speech data realizing the phone.
2. Classify (cluster) this speech into appropriately distinct categories, thus obtaining the set of allophones as well as their characterization.
3. Find for each allophone the phonetic contexts that result in the realization of the phone by that allophone.

This then is our aim. We will not be able to proceed in the above sequence, but we will achieve the desired results. It should be stressed that we are aiming for speaker independence, so the training speech will have to come from many speakers.

12.4 Collecting Realizations of Particular Phones

To start with, speech must be segmented and associated with the phonetic context:

1. Record multispeaker speech and transcribe it. (The speech may result from reading some text.)
2. Signal-process the speech, obtaining a string of speech feature vectors $z_1 z_2 \ldots, z_i \ldots$.
3. Partition the vector space into appropriate regions and label the vectors z_i by the region a_i to which they belong. Thus use vector quantization (section 1.5) to obtain the usual label strings $a_1 a_2 \ldots a_i \ldots$.
4. Using a standard phonetic alphabet and a phonetic dictionary, create phonetic word HMMs by concatenating and training elementary phone HMMs.[4]
5. Use the Viterbi algorithm to align the label string $a_1 a_2 \ldots a_i \ldots$ (and therefore also the feature vector string $z_1 z_2 \ldots z_i \ldots$) with phones. This is possible because the speech is transcribed and so can be modeled as a

---

4. See section 3.2.

large concatenation of elementary phone HMMs whose identity is determined by the base forms corresponding to the words spoken.

6. For each instance of a phone φ and its realization segment $a_1 a_2 \ldots a_k$ (the value of k differs from realization to realization), record the mapping

$$\alpha \varphi \beta \rightarrow a_1 a_2 \ldots a_k; z_1, z_2 \ldots z_k \tag{1}$$

Here $\alpha = \varphi_{-1} \ldots \varphi_{-5}$ denotes the preceding context and $\beta = \varphi_1 \ldots \varphi_5$ the following context in which the phone φ was realized.

The mappings (1) will be the basis for our derivation of allophones.

12.5   A Direct Method

We will now discuss a direct method for finding the necessary allophones. The general idea is to use decision tree questions to split the mappings into subsets corresponding to equivalence classes $\Theta_\varphi(\mathbf{h})$ of the context history $\mathbf{h} = \alpha, \beta$ of the phone φ. These classes then *are* the allophones.[5]

The data belonging to a particular context class $\Theta_\varphi(\mathbf{h})$ will be a collection of feature vectors

$$\{z_1^1, z_2^1, \ldots, z_{k_1}^1\}, \{z_1^2, z_2^2, \ldots, z_{k_2}^2\}, \ldots, \{z_1^m, z_2^m, \ldots, z_{k_m}^m\} \tag{2}$$

and corresponding label strings

$$a_1^1 a_2^1 \ldots a_{k_1}^1, a_1^2 a_2^2 \ldots a_{k_2}^2, \ldots, a_1^m a_2^m \ldots a_{k_m}^m \tag{3}$$

where we assume that exactly m contexts belonged to the equivalence class $\Theta_\varphi(\mathbf{h})$ in the mapping collection (1).[6] In this section we will deal with the feature vectors (2). Section 12.6 will justify the importance of keeping track of the label strings.

---

5. In this section the alignments are with phones, and so allophones will be produced. We can of course align with any building blocks of the overall HMM, for instance with particular parts of the phone HMM, such as transitions or states. Here we deal with HMMs that have discrete outputs associated with transitions. Only a slight change is needed for HMMs with continuous Gaussian outputs associated with states. Section 12.7 will help the reader carry out the required modifications. There, a potentially different equivalence classification $\Theta_{\varphi,s}(\mathbf{h})$ is achieved for different states s of the HMM. The analogue for the present section would be to carry out a different equivalence classification for different output producing transitions t.

6. It will soon become apparent why we deal with sets $\{z_1^j, z_2^j, \ldots, z_{k_j}^j\}$ and not with the sequences $z_1^j z_2^j \ldots z_{k_j}^j$ from which they are derived.

The basic idea is based on our belief that $\Theta_\varphi(\mathbf{h})$ will be a good equivalence class if the collection (2) is statistically pure, that is, if its elements z_i^j come from the same distribution. Hypothesizing that this distribution is Gaussian, we can find the best mean and variance and test the hypothesis that all vectors of the sets (2) were generated by that Gaussian. We will split the equivalence class $\Theta_\varphi(\mathbf{h})$ if doing so will result in significantly purer equivalence classes.

The questions of the decision tree will be addressed to phones in particular positions of the context history $\mathbf{h}$, and they will concern membership of those phones in the usual phonetic classes such as FRONT VOWEL, CONSONANT, LABIAL, FRICATIVE, STOP, PARTICULAR PHONE, and so forth.

The stopping criterion will again be based on a check set arranged on the *same* decision tree as that being developed, and using the *same* Gaussian density (as that for the development set) to evaluate the various hypothesized distributions. The last point is crucial. The Gaussian's means and variances will be determined from the development set data and will be used to compute probabilities of *both* the development set data and check set data belonging to corresponding decision tree leaves.

Let us next introduce the concept of a *Gaussian distribution belonging to a feature vector set* $\{z_1, z_2, \ldots, z_n\}$.[7] It is the distribution whose mean is the arithmetic average of the elements of the set $\{z_1, z_2, \ldots, z_n\}$, and whose *diagonal covariance matrix* Σ either is a function of the variances of the corresponding components of the vectors belonging to the set $\{z_1, z_2, \ldots, z_n\}$ or is fixed by experiment to have some appropriate value that is the same for all sets $\{z_1, z_2, \ldots, z_n\}$ considered. The value of the corresponding probability density is then denoted by $G(z|\theta)$ where θ refers to the parameters of the density function.

We can denote by θ_i the parameters corresponding to the development vector set of the ith leaf of the decision tree. At any given stage of its construction, the decision tree induces the estimate of conditional entropy

$$H_d \doteq -\sum_i \frac{N_i}{N} \sum_{z \in \mathscr{S}_i} \frac{1}{N_i} \log G(z|\theta_i) \tag{4}$$

for the feature vector data aligned with the phone φ. In (4) $\mathscr{S}_i$ denotes the

---

7. For a particular equivalence class, $\Theta_\varphi(\mathbf{h})$ is the total collection of vectors (2). Thus $n = k_1 + k_2 + \cdots + k_m$.

feature vector set associated with the ith leaf, N_i is the number of feature vectors belonging to $\mathscr{S}_i$, and $N = \sum_i N_i$.

Similarly, the entropy estimate for the check set is

$$H_c \doteq -\sum_i \frac{K_i}{K} \sum_{z \in \mathscr{S}_i'} \frac{1}{K_i} \log G(z|\theta_i) \tag{5}$$

where $\mathscr{S}_i'$ is the set of feature vectors belonging to the leaf of the check tree that corresponds to $\mathscr{S}_i$ in the development tree, K_i denotes the size of $\mathscr{S}_i'$, and $K = \sum_i K_i$.[8]

In the following development of the decision tree we will use (4) as the goodness criterion. Here is the general step based on the current state of the equivalence classification $\Theta_\varphi(\mathbf{h})$ of the mapping data (1) associated with a particular phone φ whose allophones we are trying to determine:

1. At any stage of its construction, the decision tree has leaves, indexed by i, determined by answers to questions concerning context history $\mathbf{h}$. The tree's structure and questions apply to both development and check data. The development tree's leaves correspond to feature vector sets $\{z_1, z_2, \ldots, z_n\}$, which in turn determine the appropriate densities $G(z|\theta_i)$ used also on the check sets. The latter, in turn, induce overall development and check entropy values H_d and H_c.

2. For each leaf j we find that question Q_j that will result in a minimal value of the overall development entropy, $\mathbf{H}_d(j)$ when the leaf is split. Note that the parameters θ_i and N_i in formula (4) stay the same for $i \neq j$, and that new parameters $\theta_j', \theta_j'', N_j'$, and N_j'' are tentatively determined in the node-splitting process.

3. Define

$$\hat{j} = \arg \min_j H_d(j)$$

4. For the question $Q_{\hat{j}}$ find the corresponding overall check entropy $H_c(\hat{j})$. Note that the densities $G(z|\theta_i)$ are the same for development and check sets, but the memberships of the sets $\mathscr{S}_i$ and $\mathscr{S}_i'$, and therefore the sizes N_i and K_i, are different.

---

8. Evaluation metrics different from entropy are possible, for instance, that described in [5].

5. If $H_c - H_c(\hat{j}) > \delta$, accept the split of the $\hat{j}^{th}$ leaf caused by question $Q_{\hat{j}}$, and continue the procedure. Else stop: We now have our allophones.[9]

It may be worthwhile to stress that at any stage of its development, decision tree leaves i are associated with mapping sets

$$\alpha\varphi\beta \rightarrow a_1 a_2 \ldots a_k; z_1 z_2 \ldots z_k$$

(see equation (1)) specified by histories $\mathbf{h} = \alpha, \beta$ which answer the decision tree questions in a way that will lead to the ith leaf. The sets $\mathscr{S}_i$ on which the parameters θ_i are based consist of all the feature vectors attached to the ith leaf.

12.6 The Consequences

In section 12.5 we showed how to develop for each phone φ a decision tree asking questions about the context $\mathbf{h} = \alpha, \beta$ in which the phone finds itself. The collection of leaves of the tree then constitutes the totality of allophones of the phone φ. Thus we have a means of taking any concatenation of phone base forms and transforming them into a string of allophones by simply applying to the context of each base form phone in turn the questions associated with its decision tree of the phone until a leaf is reached, and then replacing the phone by the allophone corresponding to that leaf. We therefore have a means to carry out the transformation

$$\varphi_1\varphi_2 \ldots \varphi_i \ldots \rightarrow \varphi_{j_1}^* \varphi_{j_2}^* \ldots \varphi_{j_i}^* \ldots \tag{6}$$

where $\varphi_{j_i}^*$ denotes the allophone of φ_i appropriate to the context $\alpha = \varphi_{i-1} \ldots \varphi_{i-5}, \beta = \varphi_{i+1} \ldots \varphi_{i+5}$.

The transformation (6) generates an overall HMM for $\varphi_1\varphi_2 \ldots \varphi_i \ldots$ by concatenating the elementary HMMs corresponding to the allophones $\varphi_{j_i}^*$ for $i = 1, 2, \ldots$. The straightforward way to create such elementary HMMs is to decide on their structure and train them. However, the development of the preceding section allows for a particularly accurate definition of such HMMs [5]. In fact, notice that to each leaf of the deci-

---

9. As we pointed out in section 10.5, this stopping rule may be too stringent. An alternative, worth exploring is:

For each best question Q_j at a splittable leaf, see if $H_c - H_c(j) > \delta$. If so, split the leaf (and create two new splittable leaves). If not, the jth leaf becomes unsplittable. Tree development ends when all leaves are unsplittable.

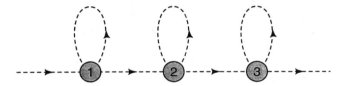

Figure 12.1
Structure of a triphone building block HMM with real vector outputs associated
with states

sion tree there corresponds a set of label strings

$$a_1^1 a_2^1 \ldots a_{k_1}^1, a_1^2 a_2^2 \ldots a_{k_2}^2, \ldots, a_1^m a_2^m \ldots a_{k_m}^m$$

that are the *realizations* of the particular allophone found in the training
data. One can then use these m label strings to construct a fenonic base-
form (section 3.6 and following) that can be used to specify the elementary HMM appropriate for that allophone. These HMMs can be trained
with the help of new training data. The word base forms obtained in this
way are commonly referred to as *leafemic base forms*.

12.7   Back to Triphones

Obviously, the method just derived can be used to cluster triphones simply
by limiting the decision tree questions to the identity of the two neigh-
boring phones, that is, by letting $\alpha = \varphi_{-1}$ and $\beta = \varphi_1$.[10] As we pointed out
in sections 9.4 and 9.6, however current state-of-the-art systems use
HMMs with real vector outputs generated by states (and not by transitions). Figure 12.1 shows the usual structure of these HMMs.

We see that this model is specified by its transition probabilities and by
three output densities, each associated with one of the states. These densities are in general mixtures of Gaussians. To simplify this section's discussion we will assume that the densities are limited to only one Gaussian
and that the training problem is to find the Gaussians' mean vectors and
covariance matrices.

Note next that logically the three model states of a triphone $\varphi_{-1} \varphi \varphi_1$ are
differently influenced by the context. In particular, the first state would be

10. In particular, there is nothing to prevent fenonic base forms from specifying
the triphone models.

more influenced by φ_{-1} than by φ_1 and the third more by φ_1 than by φ_{-1}. So naturally the equivalence classification underlying the specification of triphone HMMs should be different for their three states [2]. To each phone φ there will correspond three decision trees, one for each of the model's states. The leaves of these trees will specify (a) the mean vectors and covariance matrices of the Gaussians associated with the state to which the tree corresponds, and (b) the transition probabilities leaving the state.

The following sketch of the required algorithm will draw on the reader's understanding of the preceding sections.

1. Record multispeaker speech and transcribe it.
2. Signal-process the speech, obtaining a string of speech feature vectors $z_1 z_2 \ldots z_i \ldots$.
3. Using a standard phonetic alphabet and a phonetic base form dictionary, create phonetic word HMMs by concatenating and training elementary phone HMMs having the structure of figure 12.1.
4. Use the Viterbi algorithm to align the feature vector string $z_1 z_2 \ldots z_i \ldots$ with states of the phone models.
5. For each instance of a state-phone combination φ, s ($s \in \{1, 2, 3\}$) and its realization segment $z_1 z_2 \ldots z_k$ (the value of k differs from realization to realization), record the mappings[11]

$$\varphi_{-1} \varphi \varphi_1, s \rightarrow z_1 z_2 \ldots z_k \tag{7}$$

Define $\mathbf{h} \doteq \varphi_{-1}, \varphi_1$.
6. For each state-phone combination φ, s apply the method of section 12.5 to the mappings (7) to split the histories $\mathbf{h}$ present in the mappings into two classes, $\Theta_1(\varphi, s)$ and $\Theta_2(\varphi, s)$.[12]
7. Use this classification to specify and train triphone models. For any phone φ and for all triphone models $\varphi_{-1} \varphi \varphi_1$ whose histories $\mathbf{h}$ belong to

---

11. Note that the fact that the context of φ consists only of the immediately adjacent phones is incidental. The mappings could just as well be

$\alpha \varphi \beta, s \rightarrow z_1 z_2 \ldots z_k$

where, as in section 12.4, $\alpha = \varphi_{-1} \ldots \varphi_{-5}$ denotes the preceding context and $\beta = \varphi_1 \ldots \varphi_5$ the following context in which the phone φ and its state were realized. Limiting the context to neighboring phones in no way affects the overall method.

12. Note that the pairs of sets $\Theta_1(\varphi, s)$, $\Theta_2(\varphi, s)$ will contain different histories $\mathbf{h}$ for the different states $s \in \{1, 2, 3\}$.

the same class $\Theta_i(\varphi, s)(i = 1, 2)$, the Gaussian densities of the state s and the transition probabilities leaving it will be tied.

8. Use the Viterbi algorithm to align the feature vector string $z_1 z_2 \ldots z_i \ldots$ with states of these newly trained triphone models.

9. For each different state equivalence class $\Theta_i(\varphi, s)(i = 1, 2; s \in \{1, 2, 3\})$ record the mappings

$$\varphi_{-1} \varphi \varphi_1, s \longrightarrow z_1 z_2 \ldots z_k \tag{8}$$

where $\mathbf{h} = \varphi_{-1}, \varphi_1, \mathbf{h} \in \Theta_i(\varphi, s)$, and $z_1 z_2 \ldots z_k$ has been aligned with the state s in the triphone model designated by $\varphi_{-1} \varphi \varphi_1$.

10. Apply the method of section 12.5 to each separate mapping (8) to split the histories $\mathbf{h}$ present in the mappings into two classes.

11. The pattern is now set. We retrain the new, more refined trigram models, Viterbi-align them, construct mappings, and split histories further, continuing in this vein until the stopping criterion is invoked.

In practice, the above algorithm is carried out with many shortcuts. In particular, it is not found necessary to retrain new triphone models after every split, as is done in step 7. The "old" alignments (7) are sufficient to provide mappings (8) for many splits before new training must be undertaken.

12.8 Additional Reading

To simplify it, the general method leading to leafemes described in sections 12.3 through 12.6 was based on tree construction with the usual entropy optimization criterion (4) whose underlying distribution was given in (5). In the actual work of Gopalakrishnan and co-workers [5], however, the purity measure used relies on a particular Poisson distribution appropriate to the problem.

As pointed out in section 12.6, state-of-the-art triphone methods cluster individual states of the common HMM structure of figure 12.1. In reality, this is a considerably more complex procedure than the one sketched here. The interested reader should consult the articles of Young et al. [2] [6].

The efficacy of the methods presented in this chapter depends crucially on the accuracy of the underlying phonology, which we have here assumed to consist of a lexicon of base forms. The classical work on phonology for speech recognition is contained in reference [7]. It has been lately recognized that the base form lexicon approach to phonology is inadequate, since it does not take into account any speech dynamics, such as rate, emphasis, prosody, speech mode (conversation, reading, etc.). Currently, considerable effort is being spent in coming up with improvements [8].

References

[1] P. Ladefoged, *A Course in Phonetics*, Harcourt Brace Jovanovich, New York, 1975.

[2] S.J. Young and P.C. Woodland, "State clustering in HMM-based continuous speech recognition," *Computer Speech and Language*, vol. 8, no. 4, pp. 369–94, 1994.

[3] R. Schwartz, Y.L. Chow, O. Kimbal, S. Roucos, M. Krasner, and J. Makhoul, "Context-dependent modeling for acoustic-phonetic recognition of continuous speech," *Proceedings of IEEE International Conference on Acoustics, Speech, and Signal Processing*, pp. 1205–08, Tampa, FL, March 1985.

[4] K.F. Lee, "Context-dependent phonetic hidden Markov models for speaker-independent continuous speech recognition," *IEEE Transactions on Acoustics, Speech, and Signal Processing*, vol. ASSP-38, pp. 599–609, April 1990.

[5] L.R. Bahl, P.V. de Souza, P.S. Gopalakrishnan, D. Nahamoo, and M.A. Picheny, "Decision trees for phonological rules in continuous speech," *Proceedings of the 1991 International Conference on Acoustics, Speech, and Signal Processing*, Toronto, Canada, May 1991.

[6] S.J. Young, J.J. Odell, P.C. Woodland, "Tree-based state tying for high accuracy acoustic modelling," *Proceedings of the Human Language Technology Workshop*, pp. 307–12, Plainsboro, NJ, March 1994.

[7] P.S. Cohen and R.L. Mercer, "The phonological component of an automatic speech-recognition system," in *Speech Recognition*, D.R. Reddy, ed., pp. 275–320, Academic Press, New York, 1975.

[8] M.D. Riley and A. Ljolje, "Automatic generation of detailed pronunciation lexicons," in *Automatic Speech and Speaker Recognition*, C-H. Lee, F.K. Soong, and K.K. Paliwal, eds., pp. 285–302, Klewer Academic Publishers, Norwell, MA, 1996.

Chapter 13

Maximum Entropy Probability Estimation and Language Models

13.1 Outline of the Maximum Entropy Approach

In chapters 4 and 10 we have taken it as self-evident that it is necessary to approximate the desired language model probability $P(w|\mathbf{h})$ by $P(w|\Phi(\mathbf{h}))$, where $\Phi(\mathbf{h})$ denotes the equivalence class to which the history $\mathbf{h}$ belongs. This will fulfill three related requirements:

1. to have fewer parameters to estimate,
2. so that available data will be sufficient for the estimation, and
3. so that the probability can be constructed in a timely manner at recognition time from parameter values occupying limited storage.

In this chapter we will address directly the two first requirements, and the solution will lead automatically to the satisfaction of the third.

The idea is to construct a joint probability $P(w, \mathbf{h})$ by insisting that

- $P(w, \mathbf{h})$ should satisfy certain linear constraints, and
- $P(w, \mathbf{h})$ should organize itself in all other respects in accordance with our ignorance about everything these constraints do not specify.

As in many previous chapters, we will introduce our approach with the help of concrete examples (which is one reason we chose language modeling as our application).[1] To develop our intuition, we will start with relatively simple models and generalize later. We will use calculus to make our solutions plausible but prove no theorems about their existence or about algorithmic convergence to them. The interested reader will find proofs in the references (e.g., [1] and [2]).

---

1. At this time, although the method is completely general, the only speech recognition modules that have benefited from it are language models.

13.2   The Main Idea

Consider the task of constructing a trivariable probability $P(x, y, z)$ when knowing some of its marginals, such as $P(x, y)$ and $P(x, z)$. (We may as well continue to think of all variables as words belonging to some vocabulary.) In this case $P(x, y, z)$ would have to satisfy the constraints

$$\sum_{x,y,z} P(x, y, z)k(x, y, z|x', y') = P(x', y') \tag{1}$$

and

$$\sum_{x,y,z} P(x, y, z)k(x, y, z|x', z') = P(x', z') \tag{2}$$

where

$$k(x, y, z|x', y') \doteq \begin{cases} 1 & \text{if } x = x' \text{ and } y = y' \\ 0 & \text{otherwise} \end{cases} \tag{3}$$

and

$$k(x, y, z|x', z') \doteq \begin{cases} 1 & \text{if } x = x' \text{ and } z = z' \\ 0 & \text{otherwise} \end{cases} \tag{4}$$

If we take the view that our only knowledge about $P(x, y, z)$ is embedded in the constraints (2), then the natural approach reflecting our presumed ignorance would be to construct $P(x, y, z)$ so that of all probabilities satisfying (2), $P(x, y, z)$ would be the one inducing maximal entropy. Actually, without making the problem more difficult, instead of maximizing the entropy, we can choose P to diverge minimally from some other known probability distribution Q, that is, to minimize the divergence function[2] introduced in section 7.3. [3]

$$D(\mathbf{P}\|\mathbf{Q}) \doteq \sum_{x,y,z} P(x, y, z) \log \frac{P(x, y, z)}{Q(x, y, z)}$$

It is clear that when Q is chosen to be uniform, the divergence D is equal to the negative of entropy H plus a constant, and in that case minimizing D means maximizing H.

---

2. This generalization will be found useful in the next chapter.

13.3   The General Solution

To make the notation more compact, let $\mathbf{x} = x_1, x_2, \ldots, x_n$ denote a sequence of n random variables. Denote by $k(\mathbf{x}|i)$ the ith *constraint function* which need not be (but often is) an indicator function (as were (3) and (4)). Then the problem is:

Determine $P(\mathbf{x})$ so that

- it satisfies

$$\sum_{\mathbf{x}} P(\mathbf{x})k(\mathbf{x}|i) = d(i) \tag{5}$$

for given *constraint targets* $d(i)$, $i = 1, 2, \ldots, m$, and
- the divergence $D(\mathbf{P}\|\mathbf{Q})$ is minimal for the specified distribution $Q(\mathbf{x})$.

For the problem to make sense, the constraints (5) must be consistent, that is, there must actually exist a probability $P(\mathbf{x})$ that satisfies all the constraints simultaneously.

Note that if $k(\mathbf{x}|i)$ is an *indicator function* for the ith constraint (which is equal to 1 if the point $\mathbf{x}$ belons to the ith constraint and to 0 otherwise), then the constraints (5) simply specify certain marginals of the distribution $P(\ )$. To ensure that $P(\ )$ turns out to be a probability distribution, we must add the 0th constraint function

$$k(\mathbf{x}|0) = 1 \qquad \text{for all } \mathbf{x} \tag{6}$$

with the constraint target $d(0) = 1$.

We will find the form of the solution by the method of *undetermined Lagrangian multipliers* [4]. Therefore, for all values of $\mathbf{x}$, we set to 0 the partial derivatives with respect to $P(\mathbf{x})$ of

$$D(\mathbf{P}\|\mathbf{Q}) - \sum_i \lambda_i \left[\sum_{\mathbf{x}'} P(\mathbf{x}')k(\mathbf{x}'|i) - d(i) \right]$$

The result is

$$\log\left[\frac{P(\mathbf{x})}{Q(\mathbf{x})}\right] + 1 = \sum_{i=1}^{m} \lambda_i k(\mathbf{x}|i) + \lambda_0 \qquad \text{for all } \mathbf{x}$$

After exponentiation we then get the desired solution

$$P(\mathbf{x}) = Q(\mathbf{x})[\exp \lambda_0]\left[\exp\left\{\sum_i \lambda_i k(\mathbf{x}|i)\right\}\right]$$

where the undetermined multipliers λ_i must be chosen to satisfy the constraints (5) and (6). That is,

$$[\exp \lambda_0] \sum_{\mathbf{x}} Q(\mathbf{x}) \left[\exp\left\{ \sum_i \lambda_i k(\mathbf{x}|i) \right\} \right] k(\mathbf{x}|j) = d(j) \quad \text{for } j = 0, 1, \ldots, m$$

Reverting to the original multivariate notation, the general solution then is

$$P(x_1, \ldots, x_n) = Q(x_1, \ldots, x_n)[\exp \lambda_0] \left[\exp\left\{ \sum_i \lambda_i k(x_1, \ldots, x_n|i) \right\} \right] \quad (7)$$

with constraints

$$[\exp \lambda_0] \sum_{x_1, \ldots, x_n} Q(x_1, \ldots, x_n) \left[\exp\left\{ \sum_i \lambda_i k(x_1, \ldots, x_n|i) \right\} \right] k(x_1, \ldots, x_n|j)$$
$$= d(j) \qquad \text{for } j = 0, 1, \ldots, m \qquad \qquad \qquad (8)$$

We see from (7) that the derived probability $P(x_1, \ldots, x_n)$ is equal to $Q(x_1, \ldots, x_n)$ times a product of factors, one for each constraint in which the particular argument $x_1, \ldots, x_n$ participates.[3]

13.4 The Practical Problem

The maximum entropy approach clearly requires answers to the following two questions:

1. How do we choose the constraints? and
2. How do we solve for the parameters λ_i?

There are at least three methods for finding λ_i, all of which start with an initial guess:

1. hill climbing (*conjugate gradient descent*),
2. *iterative projection*, [5] and
3. *iterative scaling*. [1]

The last method[4] is a particular version of the second. The latter can be shown to converge [5] and, although not the most efficient one, is in principle very simple and goes as follows:

---

3. That is, the jth factor is present in the product if $k(x_1, \ldots, x_n|j) \neq 0$.
4. See section 13.8.

1. Guess at the values of λ_i, $i = 1, 2, \ldots, m$.
2. For $j = 0$ to m, do:
- keeping λ_i, $i \neq j$ fixed, find λ_j^* so as to satisfy the jth constraint;
- set $\lambda_j = \lambda_j^*$;
- end.
3. If all the constraints are sufficiently satisfied, then stop. Else go to 2.

As examples in the following sections will show, step 2 is easy to carry out provided the constraint functions $k(\mathbf{x}|j)$ can take on only the values 0 or 1. (Actually, any constant can replace the value 1.)

The problem with the iterative projection method is that it may converge slowly, particularly if the number of constraints m is large. For instance, in the illustrative problem of section 13.2, $P(x, y, z)$ was to satisfy $1 + 2N^2$ constraints, where N is the size of the alphabet (i.e., of the vocabulary when x, y, and z denote words).

The specification of constraints is a difficult problem involving two choices:

1. the choice of constraint functions $k(\mathbf{x}|j)$, that is, of the type of desirable constraints, and
2. the choice of constraint targets $d(j)$.

Fortunately, in some situations (such as the multiple decision tree problem treated in section 13.11), the needed $k(\mathbf{x}|j)$ reveal themselves naturally.

The requirement for consistency complicates the choice of $d(j)$. In the illustrative problem of section 13.2 applied to language modeling, one might wish to choose as constraints relative frequencies of bigrams observed in training data, that is, set $P(x, y) = f(x, y)$ and $P(x, z) = f(x, z)$. That would surely be consistent, because the constraints have at least one solution: the trigram relative frequencies observed in the same data. However, unless the text data were really huge, many possible bigrams x', y' would be absent from it and would have as an estimated probability $P(x', y') = 0$, surely not a desirable outcome. So constraining all the bigram probabilities to be equal to relative frequencies is not a good idea after all.

We would perhaps wish to eliminate this dilemma by imposing constraints such as

$$P(x, y) = (1 - \gamma)f(x, y) + \gamma f(x)f(y)$$

for some appropriately chosen value of γ. However, we might then run into the problem of possible lack of consistency of the resulting $P(x, y)$ and $P(x, z)$ with any probability distribution $P(x, y, z)$. Fortunately, we are not in a hopeless situation, as we will see in section 13.5.

13.5 An Example

To see concretely what is involved, let us consider the simple example of constructing a bigram language model, that is, the probability $P(x, y)$ [6]. Let it satisfy the following constraints (as usual, $C(\ )$ denotes the count function):

$$P(x, y) = f(x, y) \quad \text{if } C(x, y) \geq K$$
$$P(x) = f(x) \quad \text{if } C(x) \geq L \tag{9}$$
$$P(y) = f(y) \quad \text{if } C(y) \geq L$$
$$\sum_{x,y} P(x, y) = 1$$

It may be worthwhile to develop our formulas in detail for this first serious example. To satisfy (9), we will use three types of constraint functions whose indexing scheme is apparent:

- for all pairs x', y' such that $C(x', y') \geq K$, define

$$k(x, y | x', y') \doteq \begin{cases} 1 & \text{if } x = x', y = y' \\ 0 & \text{otherwise} \end{cases}$$

- for all x' such that $C(x') \geq L$, define

$$k_1(x, y | x') \doteq \begin{cases} 1 & \text{if } x = x' \\ 0 & \text{otherwise} \end{cases}$$

- for all y' such that $C(y') \geq L$, define

$$k_2(x, y | y') \doteq \begin{cases} 1 & \text{if } y = y' \\ 0 & \text{otherwise} \end{cases}$$

From (7) we see that the probability $P(x, y)$ will consist of a product of terms, some of which will have the form $[\exp\{\lambda_{x',y'} k(x, y | x', y')\}]$. (Other, similar terms involving the constraint functions k_1 and k_2 will also be present.) Each of these terms will be equal to 1 (i.e., its exponent will equal 0) except possibly the term with the index $x' = x$, $y' = y$. That term will be equal to something other than 1 *only if* $C(x', y') \geq K$, in which case the term will have the value

$$g(x', y') \doteq \exp\{\lambda_{x'y'}\} \tag{10}$$

It follows that if we define

$$k(x, y) = \begin{cases} 1 & \text{if } C(x, y) \geq K \\ 0 & \text{otherwise} \end{cases} \tag{11}$$

then the terms $[\exp\{\lambda_{x'y'}k(x, y|x', y')\}]$ in the expression for $P(x, y)$ can all be replaced by the single term

$$g(x, y)^{k(x,y)}$$

Similarly, terms $[\exp\{\lambda_{x'}k_1(x, y|x')\}]$ in the expression for $P(x, y)$ can all be replaced by the term $g_1(x)^{k_1(x)}$, and the terms $[\exp\{\lambda_{y'}k_2(x, y|y')\}]$ by the single term $g_2(y)^{k_2(y)}$, where[5]

$$k_1(x) = \begin{cases} 1 & \text{if } C(x) \geq L \\ 0 & \text{otherwise} \end{cases} \tag{12}$$

$$k_2(y) = \begin{cases} 1 & \text{if } C(y) \geq L \\ 0 & \text{otherwise} \end{cases} \tag{13}$$

Thus the complete expression for $P(x, y)$ becomes[6]

$$P(x, y) = g_0 \, g(x, y)^{k(x,y)} g_1(x)^{k_1(x)} g_2(y)^{k_2(y)}$$

where we have exchanged the notation of section 13.3 for one that emphasizes the product nature of maximal entropy probabilities.

Iterative projection (see section 13.4) for this example then involves the following steps:

1. Guess at initial values of $g(x, y)$, $g_1(x)$, and $g_2(y)$.
2. Set

$$g_0 = \left[\sum_{x,y} g(x, y)^{k(x,y)} g_1(x)^{k_1(x)} g_2(y)^{k_2(y)} \right]^{-1}$$

3. For all x, y such that $k(x, y) = 1$, set

$$g(x, y) = f(x, y) \left[g_0 \, g_1(x)^{k_1(x)} g_2(y)^{k_2(y)} \right]^{-1}$$

5. Again, necessarily, $g_1(x) > 0$ and $g_2(y) > 0$ for all x and y.
6. The term g_0 is a normalizing constant assuring that $\sum_{x,y} P(x, y) = 1$.

4. For all x such that $k_1(x) = 1$, set

$$g_1(x) = f(x) \left[g_0 \sum_y g(x,y)^{k(x,y)} g_2(y)^{k_2(y)} \right]^{-1}$$

5. For all y such that $k_2(y) = 1$, set

$$g_2(y) = f(y) \left[g_0 \sum_x g(x,y)^{k(x,y)} g_1(x)^{k_1(x)} \right]^{-1}$$

6. If all constraints are now sufficiently satisfied, then stop; if not, return to step 2.

Examining carefully the above steps we can conclude that the amount of computation per iteration will be essentially proportional to the amount of the training data,[7] a fact which we must be prepared to tolerate. We will reexamine computational requirements in section 13.6, where we consider a trigram model.

7. In fact.

$$\sum_{x,y} g(x,y)^{k(x,y)} g_1(x)^{k_1(x)} g_2(y)^{k_2(y)}$$

$$= \sum_{x,y:k(x,y)=1} g(x,y) g_1(x)^{k_1(x)} g_2(y)^{k_2(y)}$$

$$+ \sum_x g_1(x)^{k_1(x)} \left[\sum_y g_2(y)^{k_2(y)} - \sum_{y:k(x,y)=1} g_2(y)^{k_2(y)} \right]$$

$$= \sum_{x,y:k(x,y)=1} [g(x,y) - 1] g_1(x)^{k_1(x)} g_2(y)^{k_2(y)} + \left[\sum_x g_1(x)^{k_1(x)} \right] \left[\sum_y g_2(y)^{k_2(y)} \right]$$

The first sum has at most as many elements as there are pairs x, y in the data. This is also the amount of computation needed for step 3.

Similarly, the sum in step 4 is equal to

$$\sum_y g(x,y)^{k(x,y)} g_2(y)^{k_2(y)} = \sum_{y:k(x,y)=1} g(x,y) g_2(y)^{k_2(y)} + \sum_{y:k(x,y)=0} g_2(y)^{k_2(y)}$$

$$= \sum_{y:k(x,y)=1} [g(x,y) - 1] g_2(y)^{k_2(y)} + \sum_y g_2(y)^{k_2(y)}$$

Thus the computation necessary to carry out step 4 is again about equal to the number of pairs x, y found in the data. The same is the case for step 5.

13.6 A Trigram Language Model

Suppose we replace the variable y in section 13.5 by the history variable $\mathbf{h}$ and consider the case $\mathbf{h} = w, z$, $w \in \mathscr{V}$, $z \in \mathscr{V}$, adding to (9) the natural *bigram* constraints

$$P(x, w) = f(x, w) \quad \text{if } C(x, w) \geq M$$

$$P(x, z) = f(x, z) \quad \text{if } C(x, z) \geq M$$

As a result, the totality of constraints will become

$$P(x, \mathbf{h}) = f(x, \mathbf{h}) \quad \text{if } C(x, \mathbf{h}) \geq K$$

$$P(x, w) = f(x, w) \quad \text{if } C(x, w) \geq M$$

$$P(x, z) = f(x, z) \quad \text{if } C(x, z) \geq M \tag{14}$$

$$P(x) = f(x) \quad \text{if } C(x) \geq L$$

$$P(\mathbf{h}) = f(\mathbf{h}) \quad \text{if } C(\mathbf{h}) \geq L$$

$$\sum_{x,\mathbf{h}} P(x, \mathbf{h}) = 1$$

The complete expression for $P(x, \mathbf{h})$ will then be

$$P(x, \mathbf{h}) = g_0 \, g(x, \mathbf{h})^{k(x,\mathbf{h})} g_1(x)^{k_1(x)} g_2(\mathbf{h})^{k_2(\mathbf{h})} g_3(x, w)^{k_3(x,w)} g_4(x, z)^{k_4(x,z)}$$

where the exponents are the natural indicator functions corresponding to the constraints (14). We want to consider the amount of computation required to carry out the iterations of the corresponding iterative projection.

In particular, we will have to pick g_0 to satisfy $\sum_{x,\mathbf{h}} P(x, \mathbf{h}) = 1$. This will involve carrying out the sum

$$\sum_{x,w,z} g(x, w, z)^{k(x,w,z)} g_1(x)^{k_1(x)} g_2(w, z)^{k_2(w,z)} g_3(x, w)^{k_3(x,w)} g_4(x, z)^{k_4(x,z)}$$

$$= \sum_{x,w,z:k(x,w,z)=1} [g(x, w, z) - 1] g_1(x)^{k_1(x)} g_2(w, z)^{k_2(w,z)} g_3(x, w)^{k_3(x,w)}$$

$$\times g_4(x, z)^{k_4(x,z)}$$

$$+ \sum_{x,w,z} g_1(x)^{k_1(x)} g_2(w, z)^{k_2(w,z)} g_3(x, w)^{k_3(x,w)} g_4(x, z)^{k_4(x,z)} \tag{15}$$

The first summation on the right-hand side involves no more terms than are found in the training data,[8] so we will concentrate on the second term, which can be rewritten as

$$\sum_{x,z} g_1(x)^{k_1(x)} g_4(x,z)^{k_4(x,z)} \sum_w g_3(x,w)^{k_3(x,w)} g_2(w,z)^{k_2(w,z)} \tag{16}$$

The second sum in (16) involves $|\mathcal{V}| - N(x,z) + 1$ terms where $|\mathcal{V}|$ is the size of the vocabulary and $N(x,z)$ is the number of words w which in the training text neither follow x nor precede z. Furthermore, the result of that sum is a function $h(x,z)$ whose value for every x,z combination changes from iteration to iteration. Therefore the sum

$$\sum_{x,z} g_1(x)^{k_1(x)} g_4(x,z)^{k_4(x,z)} h(x,z)$$

requires $|\mathcal{V}|^2$ additions. It follows that the normalization step (15) requires $|\mathcal{V}|^3 - |\mathcal{V}|^2 \left(\sum_{x,z} N(x,z) - 1 \right)$ additions. Clearly, for any reasonably sized training set, the last term is bounded above by $\alpha|\mathcal{V}|^3$, where the fraction $\alpha < 1$. Hence we are faced with at least $(1 - \alpha)|\mathcal{V}|^3$ operations.

The above unpleasant conclusion is valid for a trigram model based on constraints (14). If the history $\mathbf{h}$ were more than two words long, the situation would be even worse. Section 13.7 will suggest a method of limiting the cost in computation.

13.7 Limiting Computation

The Della Pietras et al. [7] have made the important observation that if one is really interested in the conditional probability $P(w|\mathbf{h})$ (as we are) rather than in the joint $P(w, \mathbf{h})$, then it is possible to limit computation by the rule:

Use as one of the constraints $P(\mathbf{h}) = \sum_w P(w, \mathbf{h}) = f(\mathbf{h})$ \hfill (17)

where f denotes the relative frequency of events in the training set. Since by definition $\sum_{\mathbf{h}} f(\mathbf{h}) = 1$, this constraint by itself assures the proper normalization of $P(w, \mathbf{h})$ and thus avoids the analogue of step 2 of the previous section. Furthermore, since $f(\mathbf{h})$ is nonzero for at most as many

---

8. In fact, $k(x, w, z) = 1$ for at most as many different trigrams as will be found in the training data, and that number is dominated by N, the number of words making up that data.

different histories **h** as there are words N in the training data, carrying out the constraint (17) involves no more than $|\mathcal{V}| \times N$ operations, an amount which for reasonably sized data sets and vocabularies is much smaller than $(1 - \alpha)|\mathcal{V}|^3$. As we will see, when constraint (17) is imposed, the upper bound $|\mathcal{V}| \times N$ applies to the computation of all other constraints as well.

Constraint (17) might appear to make it impossible to define the probability $P(w|\mathbf{h})$ for histories **h** not observed in the training data (i.e. when $f(\mathbf{h}) = 0$). We will now show that fortunately, this is not so.

Because $P(w, \mathbf{h})$ is a product of factors, one for each constraint, then

$$P(w, \mathbf{h}) = g(\mathbf{h})q(w, \mathbf{h})$$

where $g(\mathbf{h})$ is the factor arising from constraint (17) and $q(w, \mathbf{h})$ denotes the product of the remaining factors. Constraint (17) then forces

$$g(\mathbf{h}) = \frac{f(\mathbf{h})}{\sum_{w'} q(w', \mathbf{h})}$$

so that

$$P(w|\mathbf{h}) = \frac{P(w, \mathbf{h})}{P(\mathbf{h})} = \frac{g(\mathbf{h})q(w, \mathbf{h})}{f(\mathbf{h})} = \frac{q(w, \mathbf{h})}{q(w, \mathbf{h}) + \sum_{w' \neq w} q(w', \mathbf{h})}$$

will exist even for histories **h** that do not appear in the training corpus, unless $q(w, \mathbf{h}) = 0$. Even in the latter case $P(w|\mathbf{h})$ would be well defined except if $q(w', \mathbf{h}) = 0$ for all w'. But that could only happen if for every w' there existed at least one constraint involving the pair w', **h** whose constraint target would equal 0.

We will now see in the example of estimating $P(x, |\mathbf{h})$ what the computational advantage of constraint (17) amounts to. Suppose that we impose the constraints[9]

$$P(x, \mathbf{h}) = f(x, \mathbf{h}) \quad \text{if } C(x, \mathbf{h}) \geq K$$

$$P(x, w) = f(x, w) \quad \text{if } C(x, w) \geq M$$

$$P(x, z) = f(x, z) \quad \text{if } C(x, z) \geq M$$

$$P(x) = f(x) \quad \text{if } C(x) \geq L$$

$$P(\mathbf{h}) = f(\mathbf{h}) \quad \text{for all } \mathbf{h}$$

---

9. Compare with the constraints in (14).

where, as in the preceding section, $\mathbf{h} = w, z$. Then the solution has the form

$$P(x, \mathbf{h}) = g(x, \mathbf{h})^{k(x,\mathbf{h})} g_1(x)^{k_1(x)} g_2(\mathbf{h}) g_3(x, w)^{k_3(x,w)} g_4(x, z)^{k_4(x,z)}$$

and iterative projection involves the following steps:

1. Guess at initial values of $g_1(x)$, $g_2(\mathbf{h})$, $g_3(x, w)$, and $g_4(x, z)$.
2. For all x, w, z such that $k(x, w, z) = 1$, set

$$g(x, w, z) = f(x, w, z)[g_1(x)^{k_1(x)} g_2(w, z) g_3(x, w)^{k_3(x,w)} g_4(x, z)^{k_4(x,z)}]^{-1}$$

3. For all x such that $k_1(x) = 1$, set

$$g_1(x) = f(x) \left[\sum_{w,z} g(x, w, z)^{k(x,w,z)} g_2(w, z) g_3(x, w)^{k_3(x,w)} g_4(x, z)^{k_4(x,z)} \right]^{-1}$$

4. For all w, z, set

$$g_2(w, z)$$
$$= f(w, z) \left[\sum_{x} g(x, w, z)^{k(x,w,z)} g_1(x)^{k_1(x)} g_3(x, w)^{k_3(x,w)} g_4(x, z)^{k_4(x,z)} \right]^{-1}$$

5. For all x, w such that $k_3(x, w) = 1$, set

$$g_3(x, w) = f(x, w) \left[\sum_{z} g(x, w, z)^{k(x,w,z)} g_1(x)^{k_1(x)} g_2(w, z) g_4(x, z)^{k_4(x,z)} \right]^{-1}$$

6. For all x, z such that $k_4(x, z) = 1$, set

$$g_4(x, z) = f(x, z) \left[\sum_{w} g(x, w, z)^{k(x,w,z)} g_1(x)^{k_1(x)} g_2(w, z) g_3(x, w)^{k_3(x,w)} \right]^{-1}$$

7. If all constraints are now sufficiently satisfied, then stop; if not, return to step 2.

Since, because of step 4, $g_2(w, z) = 0$ whenever $f(w, z) = 0$, steps 3 and 4 require no more than $|\mathcal{V}| \times N$ operations. The same is true about steps 2, 5, and 6 because $f(x, w, z)$, $f(x, w)$, and $f(x, z)$, respectively, are non-zero for at most N arguments. We see that the bound $|\mathcal{V}| \times N$ holds as long as constraint (17) is imposed, regardless of the complexity or number of other constraints.

13.8   Iterative Scaling

We now present a particularly efficient version of iterative projection applicable when the constraint functions $k(\mathbf{x}|j)$ are restricted to values 0 and 1. Our aim is to find the parameters $g(i)$ of the probability distribution

$$P(\mathbf{x}) = Q(\mathbf{x})g(0) \prod_{j=1}^{m} g(j)^{k(\mathbf{x}|j)} \tag{18}$$

satisfying the prescribed constraints

$$\sum_{\mathbf{x}} P(\mathbf{x})k(\mathbf{x}|j) = d(j) \qquad j = 1, 2, \ldots, m \tag{19}$$

and, of course, $\sum_{\mathbf{x}} P(\mathbf{x}) = 1$.

Iterative scaling will result in a sequence of properly normalized probability distributions $P_n(\mathbf{x})$ such that

$$\lim_{n \to \infty} P_n(\mathbf{x}) = P(\mathbf{x}) \tag{20}$$

We start with $P_0(\mathbf{x}) = Q(\mathbf{x})$, and in a round-robin fashion create successive distributions

$$P_i(\mathbf{x}) = Q(\mathbf{x})g_i(0) \prod_{j} g_i(j)^{k(\mathbf{x}|j)}, \qquad i = 1, 2, \ldots, n-1$$

satisfying the $l(i)^{th}$ constraint, where $l(i) = i \bmod (m)$. We will now show how to get the next distribution $P_n(\mathbf{x})$ from the preceding distribution $P_{n-1}(\mathbf{x})$.

To simplify notation, in what follows let $l = l(n)$ and define the sets

$$\mathscr{S}_j \doteq \{\mathbf{x} : k(\mathbf{x}|j) = 1\} \qquad j = 1, 2, \ldots, m \tag{21}$$

Our strategy will be to let[10]

$$g_n(j) = g_{n-1}(j) \qquad \text{for } j \notin \{0, l\} \tag{22}$$

and to satisfy

$$\sum_{\mathbf{x}} P_n(\mathbf{x})k(\mathbf{x}|l) = \sum_{\mathbf{x} \in \mathscr{S}_l} P_n(\mathbf{x}) = d(l) \tag{23}$$

---

10. The strategy accords with the method of alternating minimization that we know will converge to the desirable solution.

by choosing $g_n(l)$ and $g_n(0)$ so that $P_n(\mathbf{x})$ is a probability. We thus require that

$$g_n(l)g_n(0) \sum_{\mathbf{x} \in \mathscr{S}_l} Q(\mathbf{x}) \prod_{j \neq l} g_n(j)^{k(\mathbf{x}|j)} = d(l) \tag{24}$$

Because of (22), we get for $\mathbf{x} \in \mathscr{S}_l$,

$$P_n(\mathbf{x}) = g_n(l)g_n(0)Q(\mathbf{x}) \prod_{j \neq l} g_n(j)^{k(\mathbf{x}|j)} = \frac{g_n(l)g_n(0)}{g_{n-1}(l)g_{n-1}(0)} P_{n-1}(\mathbf{x}) \tag{25}$$

and for $\mathbf{x} \notin \mathscr{S}_l$,

$$P_n(\mathbf{x}) = g_n(0)Q(\mathbf{x}) \prod_{j \neq l} g_n(j)^{k(\mathbf{x}|j)} = \frac{g_n(0)}{g_{n-1}(0)} P_{n-1}(\mathbf{x}) \tag{26}$$

We may observe from (25) and (26) that P_n is a scaled version of P_{n-1}. We must now find the values of the scaling factors, that is, of the constants $g_n(l)$ and $g_n(0)$. But from (23) and (25) we get

$$\frac{g_n(l)g_n(0)}{g_{n-1}(l)g_{n-1}(0)} \sum_{\mathbf{x} \in \mathscr{S}_l} P_{n-1}(\mathbf{x}) = \sum_{\mathbf{x} \in \mathscr{S}_l} P_n(\mathbf{x}) = d(l) \tag{27}$$

and since P_n and P_{n-1} are probabilities, we get from (26) and (23) that

$$\begin{aligned}
1 - d(l) = 1 - \sum_{\mathbf{x} \in \mathscr{S}_l} P_n(\mathbf{x}) &= \sum_{\mathbf{x} \notin \mathscr{S}_l} P_n(\mathbf{x}) = \frac{g_n(0)}{g_{n-1}(0)} \sum_{\mathbf{x} \notin \mathscr{S}_l} P_{n-1}(\mathbf{x}) \\
&= \frac{g_n(0)}{g_{n-1}(0)} \left[1 - \sum_{\mathbf{x} \in \mathscr{S}_l} P_{n-1}(\mathbf{x}) \right]
\end{aligned} \tag{28}$$

We can now use (27) and (28) to solve finally for $g_n(l)$ and $g_n(0)$:[11]

$$g_n(0) = g_{n-1}(0) \frac{1 - d(l)}{[1 - \sum_{\mathbf{x} \in \mathscr{S}_l} P_{n-1}(\mathbf{x})]} \tag{29}$$

$$g_n(l) = g_{n-1}(l) \frac{d(l)}{1 - d(l)} \frac{[1 - \sum_{\mathbf{x} \in \mathscr{S}_l} P_{n-1}(\mathbf{x})]}{\sum_{\mathbf{x} \in \mathscr{S}_l} P_{n-1}(\mathbf{x})} \tag{30}$$

11. We remind the reader of the relationship between l and m: $l = n \bmod(m)$. In (29) and (30) it is tacitly assumed that the set $\mathscr{S}_l$ contains fewer elements $\mathbf{x}$ than its complement $\overline{\mathscr{S}_l}$. Should that not be the case, those calculations will be less costly if we replace $\sum_{\mathbf{x} \in \mathscr{S}_l} P_{n-1}(\mathbf{x})$ by their equivalent $1 - \sum_{\mathbf{x} \notin \mathscr{S}_l} P_{n-1}(\mathbf{x})$.

13.9   The Problem of Finding Appropriate Constraints

In the preceding sections we have acted as if constraints were given a priori, which in most situations is not the case. (In section 13.11 we treat an exception.) Usually, we have some training data and we desire to construct a probability distribution—no more is known. So we would like a procedure that would select the best K constraints from those available.

It is possible to proceed in a greedy manner[12] if we have candidate constraint functions $k(\mathbf{x}|i)$, $i = 1, \ldots, M$, $M > K$, among which we wish to decide.

Suppose we reach a point during constraint selection at which the constraints $1, 2, \ldots, m$ $(m < K)$ are already chosen. How do we determine which of the constraints $j \in \{m + 1, \ldots, M\}$ to choose next?

Assuming that the training data are divided between a development and a check set, and using the iterative projection idea, we can proceed as follows. After selecting the first m constraints, we will have constructed the distribution

$$P_m(\mathbf{x}) = Q(\mathbf{x}) \prod_{i=1}^{m} g_i^{k(\mathbf{x}|i)}$$

whose parameters were determined by the development set. The following algorithm tests the jth constraint by temporarily evaluating the potential contribution of a parameter g_j, $j \in \{m + 1, \ldots, M\}$, when the previously selected parameters g_i are held constant. Once the apparently best j^* constraint is found, all the parameters g_i, $i = 1, 2, \ldots, m$, j^* are recalculated [7]. Thus

1. For every $j \in \{m + 1, \ldots, M\}$, find the value g_j satisfying the constraint

$$g_j \sum_{\mathbf{x}} P_m(\mathbf{x}) k(\mathbf{x}|j) = d(j)$$

12. That is, we will pick the constraints one at a time, evaluating each's effect without regard to those constraints yet to be selected. (This is analogous to our approach to decision tree construction in chapters 10 through 12.) In contrast, the "right" (but computationally not feasible) way would be to compare to each other all possible sets of constraints and then pick the best set.

and construct the probability

$$P_j^*(\mathbf{x}) = \frac{P_m(\mathbf{x})g_j^{k(\mathbf{x}|j)}}{\sum_{\mathbf{x}'} P_m(\mathbf{x}')g_j^{k(\mathbf{x}'|j)}}$$

2. Choose that value of j for which the entropy $P_j^*(\mathbf{x})$ induces is minimal.[13] Call it j^*.

3. Recalculate the values g_i, $i = 1, \ldots, m$ and g_{j^*} to satisfy all constraints, and thus obtain the next tentative probability $P_{m+1}(\mathbf{x})$.

4. If $P_{m+1}(\mathbf{x})$ also lowers the check set's entropy, then:

(a) Remove the chosen $(j^*)^{th}$ constraint from the candidate set $\{m+1, \ldots, M\}$ and place it into the accepted set $\{1, \ldots, m\}$, which now becomes the set $\{1, \ldots, m+1\}$.

(b) Set $m \rightarrow m+1$.

(c) Accept the new $P_m(\mathbf{x})$ as the basic probability.

(d) If $m < K$ then go to 1. Else go to 6.

5. If $P_{m+1}(\mathbf{x})$ does not lower the entropy of the check set, then there is no point going any further. Set $K = m$.

6. Add the check set to the development set and recalculate all parameters corresponding to the accepted constraint set $\{1, \ldots, m\}$, thus obtaining the final maximum entropy probability $P(\mathbf{x})$ that satisfies these constraints.

13.10   Weighting of Diverse Evidence: Voting

In many cases of interest there are k separate sources of knowledge with observables $y_1, y_2, \ldots, y_k$ to be used to predict the outcome x of some experiment. How do we use all the observations $y_1, y_2, \ldots, y_k$ to get the "best" estimate of x if only the conditional probabilities $p(x|y_i)$, $i = 1, 2, \ldots, k$ are known?[14]

---

13. Here we implicitly assume that the goal of the probability construction is language modeling, that is, the construction of a conditional probability $P(w|\mathbf{h})$. We are seeking to minimize the conditional entropy of $P(w|\mathbf{h})$ induced by $P(\mathbf{x})$. Of course, $\mathbf{x} = \mathbf{h}, w$. If the goal of the probability construction process is other than language modeling, we may need to use a criterion other than entropy.

14. In section 13.11 we will apply the answer to the construction of multiple decision trees.

One possible answer is to use training data to construct the probability $P(x|y_1, \ldots, y_k)$ by the maximum entropy approach [8]. That is, use constraints

$$P(y_1, y_2, \ldots y_k) = f(y_1, y_2, \ldots, y_k)$$

$$P(x, y_i) = p(x|y_i)f(y_i) \qquad i = 1, 2, \ldots, k$$

The resulting functional form is

$$P(x, y_1, \ldots, y_k) = g(y_1, \ldots, y_k) \prod_{i=1}^{k} g_i(x, y_i)$$

and the desired probability then has the form

$$P(x|y_1, \ldots, y_k) = \frac{\prod_{i=1}^{k} g_i(x, y_i)}{\sum_{x'} \prod_{i=1}^{k} g_i(x', y_i)}$$

which, once the g_i functions are known, is not costly to compute at run time if the x-space is not too large.

The corresponding iterative projection algorithm is

1. Guess the initial values of $g_1(x, y_1), \ldots, g_k(x, y_k)$.
2. For all combinations of values $y_1, \ldots, y_k$ in the training data, set

$$g(y_1, \ldots, y_k) = f(y_1, \ldots, y_k) \left[\sum_{x} \prod_{i=1}^{k} g_i(x, y_i) \right]^{-1}$$

3. Set

$$g_1(x, y_1) = p(x|y_1)f(y_1) \left[\sum_{y_2, \ldots, y_k} g(y_1, \ldots, y_k) \prod_{i=2}^{k} g_i(x, y_i) \right]^{-1}$$

4. Set

$$g_2(x, y_2) = p(x|y_2)f(y_2) \left[\sum_{y_1, y_3, \ldots, y_k} g(y_1, \ldots, y_k) \prod_{i=1, i \neq 2}^{k} g_i(x, y_i) \right]^{-1}$$

5. And so forth.

None of the steps is too difficult to carry out: The sums contain only as many terms as there are values $y_1, \ldots, y_k$ found coexisting in the training data.

13.11   Limiting Data Fragmentation: Multiple Decision Trees

A major problem in decision tree design is that of fragmentation of data.[15] The ith leaf of the tree corresponds to that portion of the training data belonging to the equivalence class Ψ_i determined by the answers to questions along the path from the tree root to the ith leaf. No data is common to two different leaves; the data at all of the leaves is the totality of training data. The tree is developed by selecting questions whose quality is determined by the data residing at the leaf to be split, and the resulting split of the leaf distributes its data among the two leaves being created. Thus tree development is based on progressively less and less data.

13.11.1   Combining Different Knowledge Sources

One way to counteract data fragmentation is to limit the depth of the trees being developed by creating multiple trees, each based on the totality of data [8].[16] This can be done, for instance, by basing the different trees on questions that address different features of the data. Different trees would constitute separate knowledge sources which can be combined by voting, as discussed in section 13.10. An example appropriate to language modeling might be two trees, the first based on the history **h** being regarded as a word string, and the other determined by questions that interrogate the (partial) *parse tree*[17] pertaining to **h**.

The multiple tree approach determines naturally the required maximal entropy constraints. Let $\Psi_i(j)$ denote the equivalence class of the ith leaf of the jth tree. (Figure 13.1 illustrates the notation for two trees. In general there are M trees. The jth tree has $N(j)$ leaves.) Then the corresponding constraint functions are given by

$$k(\mathbf{h}|i, j) = \begin{cases} 1 & \text{if } \mathbf{h} \in \Psi_i(j) \\ 0 & \text{otherwise} \end{cases} \tag{31}$$

15. See Section 10.14.2.

16. In section 10.14.2, we have already suggested a relatively primitive way of multiple tree construction and use.

17. A parse tree is a standard representation of syntactic analysis of a sentence. It names the functions of the various phrases present in the sentence and establishes relationships between them. A good introductory exposition can be found in the section on phrase structure grammars, pp. 215–27 of [9].

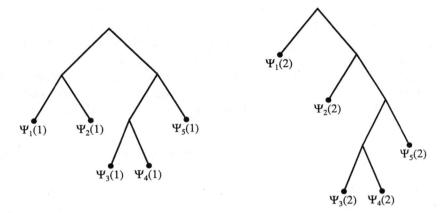

Figure 13.1
Example of two decision trees developed on the basis of the same data but different kinds of questions

and the desired probability distribution has the form

$$P(w, \mathbf{h}) = g(\mathbf{h}) \prod_{j=1}^{M} \prod_{i=1}^{N(j)} g(w, i, j)^{k(\mathbf{h}|i, j)} \tag{32}$$

and satisfies

$$\sum_{\mathbf{h}} P(w, \mathbf{h}) k(\mathbf{h}|i, j) = d(w|i, j) \sum_{\mathbf{h}} f(\mathbf{h}) k(\mathbf{h}|i, j)$$

$$j = 1, 2, \ldots, M; i = 1, 2, \ldots, N(j) \tag{33}$$

and

$$\sum_{w} P(w, \mathbf{h}) = f(\mathbf{h})$$

To assure the consistency of constraints, $d(w|i, j)$ would normally equal the relative frequency $f(w|i, j)$ of the word w at the ith leaf of the jth tree, and the factor $\sum_{\mathbf{h}} f(\mathbf{h}) k(\mathbf{h}|i, j)$ is the frequency with which the histories in the training corpus belong to the leaf equivalence class $\Psi_i(j)$.

Although (32) looks very formidable, note from (31) that for any particular history $\mathbf{h}$, there will be only one value of i in the jth tree, say $i(\mathbf{h}, j)$, for which $k(\mathbf{h}|i, j) \neq 0$. Therefore (32) can be replaced by

$$P(w, \mathbf{h}) = g(\mathbf{h}) \prod_{j=1}^{M} g(w, i(\mathbf{h}, j), j) \tag{34}$$

Thus the run time use of a multiple decision tree is very simple. At the ith leaf of the jth tree is stored the function $g(w, i, j)$. The terms needed to form the product (34) are determined by answers to the questions applied by each of the M trees to the particular history $\mathbf{h}$ of current interest.

Let us finally consider the problem of finding the values of the functions g. Denote by $\mathscr{S}(i, j)$ the set of all histories $\mathbf{h}$ in the training data that correspond to the ith leaf of the jth tree. Then $g(w, i, j)$ must satisfy

$$g(w, i, j) = d(w|i, j) \sum_{\mathbf{h} \in \mathscr{S}(i,j)} f(\mathbf{h}) \left[\sum_{\mathbf{h} \in \mathscr{S}(i,j)} g(\mathbf{h}) \prod_{j' \neq j} g(w, i(\mathbf{h}, j'), j') \right]^{-1}$$

(35)

Formulas (35) for each i, j combination together with

$$g(\mathbf{h}) = f(\mathbf{h}) \left[\sum_w \prod_{j'=1}^{M} g(w, i(\mathbf{h}, j'), j') \right]^{-1}$$

can then be used in an iterative projection algorithm to find the desired values of the factors g.

13.11.2   Spontaneous Multiple Tree Development

It is also possible to develop multiple trees as they are needed. Figure 13.2, in which we have two trees in existence and the possibility of starting a third, illustrates the basic idea [8]. We can take the view that we have three trees, the third having only one equivalence class $\Psi_1(3)$ (which is degenerate in that it contains all histories $\mathbf{h}$). We have a set of questions at our disposal, and the problem is which, if any, of the equivalence classes $\Psi_i(j), j = 1, 2, 3, i = 1, 2, \ldots, N(j)$ to split.[18]

Obviously, the evaluation criterion used to decide which $\Psi_i(j)$ to split will be based on $P(w, \mathbf{h})$ which, at the present stage of the multiple tree development, has the form (34) with $M = 3$.[19] As we evaluate candidate

18. The expression $N(j)$ denotes the number of leaves in the jth tree. Note that in the example, $N(3) = 1$.

19. This is because there are three actual trees. The third, vestigial tree, does impose a constraint, namely

$$\sum_{\mathbf{h}} P(w, \mathbf{h}) = f(w)$$

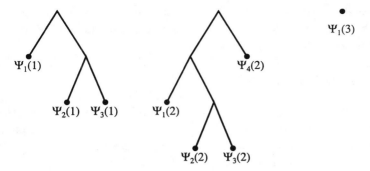

Figure 13.2
Current state of development of multiple decision trees based on the same data
and same set of potential questions

questions splitting the various equivalence classes $\Psi_i(j)$, we must not only
replace the function $g(w, i, j)$ by a new pair $g(w, i_0, j)$ and $g(w, i_1, j)$, but
we should also recalculate all the remaining old functions. This is too big
a computational burden. We must approximate.

One way to do so, as suggested in section 13.9, is to compute

$$g'(w, i_0, j) = d(w|i_0, j) \sum_{\mathbf{h} \in \mathscr{S}(i_0, j)} f(\mathbf{h}) \left[\sum_{\mathbf{h} \in \mathscr{S}(i_0, j)} g(\mathbf{h}) \prod_{j' \neq j} g(w, i(\mathbf{h}, j'), j') \right]^{-1}$$

and

$$g'(w, i_1, j) = d(w|i_1, j) \sum_{\mathbf{h} \in \mathscr{S}(i_1, j)} f(\mathbf{h}) \left[\sum_{\mathbf{h} \in \mathscr{S}(i_1, j)} g(\mathbf{h}) \prod_{j' \neq j} g(w, i(\mathbf{h}, j'), j') \right]^{-1}$$

We then use these functions to renormalize

$$g'(\mathbf{h}) = f(\mathbf{h}) \left[\sum_w \left[\prod_{j' \neq j} g(w, i(\mathbf{h}, j'), j') \right] [g'(w, i_0, j)g'(w, i_1, j)] \right]^{-1}$$

and finally to evaluate the candidate question on the basis of the probability,

$$P(w, \mathbf{h}) = \begin{cases} g'(\mathbf{h}) \prod_{j=1}^{M} g(w, i(\mathbf{h}, j), j) & \text{if } i(\mathbf{h}, j) \notin \{i_0, i_1\} \\ g'(\mathbf{h})g'(w, i_0, j) \prod_{j' \neq j} g(w, i(\mathbf{h}, j'), j') & \text{if } i(\mathbf{h}, j) = i_0 \\ g'(\mathbf{h})g'(w, i_1, j) \prod_{j' \neq j} g(w, i(\mathbf{h}, j'), j') & \text{if } i(\mathbf{h}, j) = i_1 \end{cases} \quad (36)$$

Of course, once the apparently best split (consisting of a choice of a leaf $\Psi_i(j)$ and a question) is selected, all the g functions should be recalculated to reflect the complete set of current constraints.

13.12   Remaining Unsolved Problems

In section 13.9 we showed how one might go about selecting the best K constraints from a candidate set of M constraints. Of course, one would really like to have a method that would suggest candidate constraints itself on the basis of available data. Finding such a method presents the first problem.

The second problem is that to be sure that the imposed constraints are mutually consistent, one usually limits oneself to target values $d(i)$ (see (31)) which are themselves empirical averages (such as relative frequencies) derived from the same training data from which the desired probability will be estimated. But if these targets are, for instance, marginal probabilities, then better estimates exist than relative frequencies.[20] How do we use these and preserve the mutual consistency property of all the constraints? Or will a good solution be found even when consistency is lacking?[21]

The final problem seems the most crucial. In practice, there will always be a residue of uncertainty about the value of the constraint targets $d(i)$. How are we to incorporate the *degree* of this uncertainty into the formulation?

Maximum entropy is not a panacea. It can lead to absurd results, just because of the dichotomy that is the basis of the formulation: Either we know a constraint perfectly or not at all. For instance, for sufficiently large values of an integer L, we would be willing to impose the constraint

$$P(x) = f(x) \qquad \text{if } C(x) \geq L$$

20. This subject is discussed in chapter 15 [10].

21. Some recent experimental work [11] indicates that it will. Actually, under rather broad general conditions, alternating minimization will converge [12] [13]. The distribution obtained will lead to a constraint vector (whose i^{th} component is the left-hand side of (5)) **K** whose relation to the target vector **D** (whose i^{th} component is the right-hand side of (5)) minimizes the generalized divergence

$$D(\mathbf{K}\|\mathbf{D}) \doteq \sum_i \left(k_i \log \frac{k_i}{d_i} + d_i - k_i \right)$$

The required normalization $\sum_x P(x) = 1$ then gives us the solution

$$P(x) = g_0\, g(x)^{k(x)}$$

where

$$k(x) = \begin{cases} 1 & \text{if } C(x) \geq L \\ 0 & \text{otherwise} \end{cases}$$

Unfortunately, the resulting probability has the following undesirable property:

$$P(x) = g_0 \qquad \text{for all } x \text{ such that } C(x) < L$$

Therefore, because we were noncommittal about the low count values of $P(x)$, the method punished us by insisting that all these values be equal. But we surely know that at least for some sufficiently large ε,

$$P(x) > P(x') \qquad \text{whenever } f(x) \geq f(x') + \varepsilon$$

Yet we have shown no elegant way to incorporate this conviction into our framework.

13.13 Additional Reading

Maximum entropy methods are a subject of current intensive investigation. They are of interest to a wide variety of scientists such as astronomers and biologists. Symposia dealing with the topic continue to be organized [14].

The most active investigator in the field is Csiszar. In [15] and [16] he considers the scientific arguments for the maximum entropy method. In [17] he gives a geometric interpretation to the alternating minimization method of convergence.

Section 13.4 presented iterative projection as a method in which parameters are adjusted in a round-robin fashion, one at a time. Actually, Darroch and Ratcliff [1] improve on this version of the Brown algorithm [5] by showing that the adjustment may be carried out in parallel, that is, that the set of new parameter values λ_j^*, $j = 1, \ldots, m$, may be obtained entirely on the basis of old values λ_i, $i = 1, \ldots, m$, which the former then replace. This, of course, has the great computational advantage in that only one pass through the training data is necessary for each iteration. The Della Pietras and Lafferty [18] modify this algorithm to achieve faster convergence (in fewer steps) without increasing the computational load.

Finally, Csiszar and Tusnady [2] consider the general conditions on functions whose parameters can also be optimized by the Darroch-Ratcliff procedure.[22]

Interesting applications to Bolzmann machine learning can be found in Byrne's article [19].

References

[1] J.N. Darroch and D. Ratcliff, "Generalized iterative scaling for log-linear models," *The Annals of Mathematical Statistics*, vol. 43, pp. 1470–80, 1972.

[2] I. Csiszar and G. Tusnady, "Information geometry and alternating minimization procedures," *Statistics and Decisions*, supp. iss. no. 1, pp. 205–37, R. Oldenburg Verlag, Munich, 1984.

[3] S. Kullback, *Information Theory and Statistics*, Wiley, New York, 1959.

[4] F.B. Hildebrand, *Methods of Applied Mathematics*, ch. 2, Prentice-Hall, Englewood Cliffs, NJ, 1952.

[5] D. Brown, "A note on approximations to discrete probability distributions," *Information and Control*, vol. 2, pp. 386–92, 1959.

[6] R. Lau, R. Rosenfeld, and S. Roukos, "Method for building scalable N-gram language models using maximum likelihood maximum entropy N-gram models," U.S. patent filed 1994.

[7] A.L. Berger, S.A. Della Pietra and V.J. Della Pietra, "A maximum entropy approach to natural language processing," *Computational Linguistics*, vol. 22, no. 1, pp. 39–72, March 1996.

[8] S.A. Della Pietra, V.J. Della Pietra, and R.L. Mercer: personal communication.

[9] J. Lyons, *Introduction to Theoretical Linguistics*, Cambridge University Press, London, 1968.

[10] I.J. Good, "The population frequencies of species and the estimation of population parameters," *Biometrica*, vol. 40, nos. 3 and 4, pp. 237–64, 1953.

[11] R. Rosenfeld, "A maximum entropy approach to adaptive statistical language modeling," *Computer Speech and Language*, vol. 10, no. 3, pp. 187–228, July 1996.

[12] C.L. Byrne, "Iterative image reconstruction algorithms based on cross-entropy minimization," *IEEE Transactions on Image Processing*, vol. 2, no. 1, pp. 96–103, January 1993.

[13] C.L. Byrne, "Erratum and addendum to 'Iterative image reconstruction algorithms based on cross-entropy minimization,'" *IEEE Transactions on Image Processing*, vol. 4, no. 2, pp. 226–27, February 1995.

---

22. That is, they determine what other functions besides the divergence $D(\mathbf{P}\|\mathbf{Q})$ can be so optimized.

[14] J. Skilling, ed., *Maximum Entropy and Bayesian Methods*, Kluwer Academic Publishers, Dordrecht, The Netherlands, 1988.

[15] I. Csiszar, "Why least squares and maximum entropy? An axiomatic approach to inference for linear inverse problems," *The Annals of Statistics*, vol. 19, no. 4, pp. 2032–66, 1991.

[16] I. Csiszar, "An extended maximum entropy principle and a bayesian justification," *Bayesian Statistics*, vol. 2, pp. 83–98, 1985.

[17] I. Csiszar, "A geometric interpretation of Darroch and Ratcliff's generalized iterative scaling," *The Annals of Statistics*, vol. 17, no. 3, pp. 1409–13, 1989.

[18] S.A. Della Pietra, V.J. Della Pietra, and J. Lafferty, *Inducing features of random fields*, Technical Report CMU-CS-95-144, Dept. of Computer Science, Carnegie Mellon University, Pittsburgh, PA, 1995.

[19] W. Byrne, "Alternating minimization and Bolzmann machine learning," *IEEE Transactions on Neural Networks*, vol. 4, no. 4, pp. 612–20, July 1992.

Chapter 14

Three Applications of Maximum Entropy Estimation to Language Modeling

14.1 About the Applications

In this chapter we will apply maximum entropy estimation to three problems of language modeling: adaptation to a new domain, dynamic adjustment of probabilities during the dictation of text, and creation of a cache language model. Our aim is to illustrate the capability of the maximum entropy approach in meeting various challenges. The reader will find many opportunities to use this method to his own advantage.[1]

Language model adaptation is needed when insufficient training data is provided for complete language model estimation and when the new domain is sufficiently similar to an old one (for which a language model is assumed to exist) that the main structural properties of the old may be presumed to apply to the new.

Dynamic adjustment of probabilities deals with a varied domain in which the production of words is highly nonstationary: In some parts of the test (e.g., in a particular document) a subvocabulary suddenly bursts forth only to become dormant again later. For instance, a user may dictate documents concerned with the stock market only rarely, but when he does, his speech will suddenly be replete with words like *broker, futures, Dow, hedge, option,* and the like.

The method employed to alleviate the above problem involves precomputation that determines *trigger words* whose appearance in the text affects the distribution of future words. Of course, this can be successful only if the dynamic behavior being modeled is observable somewhere in the training corpus.

---

1. Dynamic adjustment of probabilities via *triggers* has definitely merited incorporation into practical recognizers.

Finally, a speech recognizer must provide for cases where a document employs words from a heretofore unobserved subvocabulary. Then the language model dynamic adjustment must be direct, based on word frequencies observed in the current document so far. This leads to the *cache language model*.

14.2   Simple Language Model Adaptation to a New Domain

We assume here that we have a fully satisfactory language model Q derived for a relatively rich domain #1, that we wish to determine a language model P for domain #2, and that the training data provided for the latter domain is of insufficient size to allow for a reliable estimate of P that would induce a reasonable perplexity [1].[2] If the two domains are not too dissimilar, it is reasonable to estimate P so it diverges minimally from Q and satisfies additional constraints whose value can be determined from the admittedly sparse training data for domain #2. That is, we will want to find the distribution P that minimizes the divergence $D(\mathbf{P}\|\mathbf{Q})$ under imposed constraints.

As a simple example, suppose we have a satisfactory bigram language model $Q(w_1, w_2)$ for domain #1 and that we wish to estimate $P(w_1, w_2)$ for domain #2. Assuming that the data from domain #2 is adequate only for unigram constraints[3]

$$P(w_1) = f_2(w_1)$$
$$P(w_2) = f_2(w_2) \tag{1}$$

the resulting distribution will have the form (see equation (18) of chapter 13)

$$P(w_1, w_2) = Q(w_1, w_2)g_1(w_1)g_2(w_2) \tag{2}$$

with the correction factors g_1 and g_2 to be determined so that P satisfies the imposed constraints (1).

---

2. For instance, the data may be sufficient for a bigram but not for a trigram estimate. In fact, a trigram estimate would lead to a distribution
$$P(w_3|w_1, w_2) = \lambda_3 f(w_3|w_1, w_2) + \lambda_2 f(w_3|w_2) + \lambda_1 f(w_3)$$
whose interpolation weight λ_3 would be relatively low for most histories w_1, w_2.
3. We subscript the relative frequency function f_2 to stress that it is derived from domain #2.

The form (2) implies that the vocabulary of domain #2 is contained in the vocabulary of domain #1.[4] If this is not the case, then the presence of Q in the form (2) would make P assign 0 probability to words present in domain #2 but not belonging to domain #1. Therefore, before starting the probability construction process, we must broaden the argument span of Q to include all bigrams from domain #2.

Suppose words of both domains can be assigned to categories π, such as, for instance *parts of speech*. In fact, additional pseudogrammatical categories may be provided such as *street name, last name, price*, etc. Also, define sets

$$\mathscr{S}_1 \doteq \{w : w \text{ belongs to domain } \#1\}$$

$$\mathscr{S}_2 \doteq \{w : w \text{ belongs to domain } \#2 \text{ and } \mathbf{not} \text{ to domain } \#1\}$$

Then the broadened distribution Q^* can be defined by[5]

$$Q^*(w_1, w_2)$$

$$\doteq \begin{cases} f_2(\mathscr{S}_1, \mathscr{S}_1)Q(w_1, w_2) & \text{if } w_1 \in \mathscr{S}_1, w_2 \in \mathscr{S}_1 \\[2mm] f_2(\mathscr{S}_1, \mathscr{S}_2)\sum_\pi Q(w_1, \pi)\dfrac{f_2(w_2|\pi)}{f_2(\mathscr{S}_2|\pi)} & \text{if } w_1 \in \mathscr{S}_1, w_2 \in \mathscr{S}_2 \\[3mm] f_2(\mathscr{S}_2, \mathscr{S}_1)\sum_\pi \dfrac{f_2(w_1|\pi)}{f_2(\mathscr{S}_1|\pi)}Q(\pi, w_2) & \text{if } w_1 \in S_2, w_2 \in \mathscr{S}_1 \\[3mm] f_2(\mathscr{S}_2, \mathscr{S}_2)\sum_{\pi_1, \pi_2}\dfrac{f_2(w_1|\pi_1)}{f_2(\mathscr{S}_1|\pi_1)} & \\[2mm] \times Q(\pi_1, \pi_2)\dfrac{f_2(w_2|\pi_2)}{f_2(\mathscr{S}_2|\pi_2)} & \text{if } w_1 \in \mathscr{S}_2, w_2 \in \mathscr{S}_2 \end{cases} \qquad (3)$$

Q^* then replaces Q in formula (2), and we can then proceed to estimate the factors $g_1(\ )$ and $g_2(\ )$ by the methods of chapter 13.

Constraints (1) obviously are applicable only if we have full confidence that the size of the training set is sufficient to allow us to estimate probabilities using relative frequencies. Note that since the distribution $P(w_1)$

4. The constraints (1) will force P to assign probability 0 to words not in domain #2, which will remedy the fact that Q may assign positive probability to them.

5. Clearly, we define

$$f_2(\mathscr{S}_i, \mathscr{S}_j) \doteq \sum_{w_1 \in \mathscr{S}_i} \sum_{w_2 \in \mathscr{S}_j} f_2(w_1, w_2)$$

etc. This insures that the probability Q^* will be properly normalized.

does not formally imply anything about the distribution $P(w_2)$,[6] we can constrain these parameters to equal any probabilities without risking a lack of consistency. Thus, instead of (1) we can use

$$P(w_1) = \theta_1(w_1)$$

$$P(w_2) = \theta_2(w_2)$$

where θ_1 and θ_2 are any probability distributions, for instance the Good-Turing estimates [2] obtained from the training set.[7]

Disregarding the minor complications of the broadening (3) of the probability Q, we have in this section warped the distribution Q to conform it to some known properties of the distribution P. The unigram constraints (1) constitute only one of many possible such properties.

This method of adaptation that relates the probability P to a reliable probability Q can even be used in real time where domain #1 is the training data, domain #2 is the actual document being dictated, and the really sparse training data for domain #2 consists of the portion of the document that has been dictated so far. We will say more about this opportunity in section 14.5.

14.3   A More Complex Adaptation

We will now treat the general problem of adaptation from a rich to a sparse domain. As before, we will assume that the two domains are sufficiently similar that Q is an appropriate prior for P. We will denote the arguments of both distributions by $\mathbf{w} = \mathbf{h}, w$ where $\mathbf{h}$ is the history of which we take cognizance.

We will first estimate a maximum entropy model $Q(\mathbf{w})$ for the rich domain and then use $Q(\mathbf{w})$ as the prior for the $P(\mathbf{w})$ of the sparse domain. $Q(\mathbf{w})$ should be a maximum entropy model because we intend to take

---

6. That is, the fact that knowing the reality behind w_1 and w_2 we would wish $P(w_1 = v) = P(w_2 = v)$ for all $v \in \mathscr{V}$ does not prevent us from formally regarding w_1 and w_2 as two completely different random variables. In any case, regardless of the choice of $P(w_1)$ and $P(w_2)$, there will always exist a distribution $P(w_1, w_2)$ such that

$$\sum_{w_2} P(w_1, w_2) = P(w_1) \quad \text{and} \quad \sum_{w_1} P(w_1, w_2) = P(w_2)$$

because a possible distribution is simply $P(w_1, w_2) = P(w_1)P(w_2)$.

7. For this and other alternatives, see chapter 15.

advantage of the fact that $Q(\mathbf{w})$ and $P(\mathbf{w})$ can have the same structure, with most of the constraints of $P(\mathbf{w})$ being a subset of the constraints of $Q(\mathbf{w})$.

In estimating $Q(\mathbf{w})$ from the rich domain, we will impose (among others) constraints that will assure that $Q(\mathbf{w}) = 0$ for all strings $\mathbf{w}$ that contain words not belonging to the sparse domain.[8] This will remove any possible difficulties that Q might cause as a prior to P.

Let the constraint functions for the estimation of either Q or P be denoted by $k(\mathbf{w}|i)$.[9] Let the first m functions define constraints applied only to the sparse domain. Let the next n functions be common to both domains, and let the last $L - m - n$ functions define constraints applied only to the rich domain.[10] Of course, we take here the view that the first $n + m$ constraints are those for which we have enough data in the sparse text. Then, because Q was estimated by the maximum entropy method, it has the product form[11]

$$Q(\mathbf{w}) = Z_Q \prod_{i=m+1}^{n+m} e^{\lambda_i k(\mathbf{w}|i)} \prod_{j=n+m+1}^{L} e^{\lambda_j k(\mathbf{w}|j)}$$

Therefore

$$P(\mathbf{w}) = Z_P^* Q(\mathbf{w}) \prod_{i=1}^{n+m} e^{\lambda_i^* k(\mathbf{w}|i)}$$

$$= Z_P^* Z_Q \prod_{l=1}^{m} e^{\lambda_l^* g_l(\mathbf{w})} \prod_{i=m+1}^{n+m} e^{(\lambda_i + \lambda_i^*) k(\mathbf{w}|i)} \prod_{j=n+m+1}^{L} e^{\lambda_j k(\mathbf{w}|j)} \qquad (4)$$

$$= Z_P R_Q(\mathbf{w}) \prod_{i=1}^{n+m} e^{\gamma_i k(\mathbf{w}|i)}$$

8. That is particularly simple if the constraint targets are relative frequencies evaluated over the rich domain, or their sums. Then these frequencies need only to be renormalized so they sum to 1 over their arguments restricted to strings in the sparse domain.

9. As in all other examples in this book, we will limit the functions $k(\mathbf{w}|i)$, to be indicator functions of some set $\mathscr{S}_i$.

10. That is, the totality of constraints that P must satisfy are defined by $k(\mathbf{w}|i)$, $i \in \{1, 2, \ldots, m+n\}$, and those that Q must satisfy are defined by $k(\mathbf{w}|i)$, $i \in \{m+1, m+2, \ldots, L\}$.

11. $Z_Q = e^{\lambda_0}$ is the usual normalizing factor assuring that $\sum Q(\mathbf{w}) = 1$.

where

$$R_Q(\mathbf{w}) = \prod_{j=n+m+1}^{L} e^{\lambda_j k(\mathbf{w}|j)} \tag{5}$$

The structure (4) places in evidence exactly what part of the probability shape of $Q(\mathbf{w})$ is being directly accepted by $P(\mathbf{w})$ (i.e., $R_Q(\mathbf{w})$) and which part is being newly estimated (i.e., $\prod_{i=1}^{n+m} e^{\gamma_i k(\mathbf{w}|i)}$).

Here is a more detailed outline of how to proceed:

1. Inspecting the sparse domain, select n constraints for whose imposition both domains contain enough data.

2. Select m additional constraints important to the sparse domain but lacking enough support in the rich domain.

3. Find in the rich domain $L - n - m$ additional constraints that can be imposed in that domain so that it is adequately characterized by its $L - m$ constraints.

4. Impose on the rich domain constraints that will assure that $Q(\mathbf{w}) = 0$ for all strings $\mathbf{w}$ that contain words not belonging to the sparse domain.

5. Carry out the maximum entropy algorithm on the rich domain, obtaining the parameters λ_i for the total set of its $L - m$ constraints, thus deriving the $Q(\mathbf{w})$ distribution, and therefore the function $R_Q(\mathbf{w})$ defined by (5).

6. Imposing the $n + m$ constraints from the sparse domain, evaluate the corresponding parameters $\gamma_i, i = 1, 2, \ldots, n + m$ that define the final distribution

$$P(\mathbf{w}) = Z_P R_Q(\mathbf{w}) \prod_{i=1}^{n+m} e^{\gamma_i k(\mathbf{w}|i)}$$

7. Keeping parameters $\gamma_i, i = m + 1, m + 2, \ldots, m + n$ fixed, impose on the rich domain[12] only the constraints $j = m + n + 1, \ldots, L$ to obtain a new distribution

$$Q'(\mathbf{w}) = Z'_Q \prod_{i=m+1}^{n+m} e^{\gamma_i k(\mathbf{w}|i)} \prod_{j=n+m+1}^{L} e^{\lambda'_j k(\mathbf{w}|j)}$$

that is, new parameters λ'_j.

---

12. In this and the following steps we are matching the rich and sparse parameters to each other to get the best possible estimated P.

8. Finally, keeping the $\lambda'_j, j = n + 1, \ldots, L$ fixed, and defining

$$R'_Q(\mathbf{w}) = \prod_{j=n+m+1}^{L} e^{\lambda'_j k(\mathbf{w}|j)}$$

impose constraints $i = 1, \ldots, n + m$, from the sparse domain and obtain the final distribution

$$P'(\mathbf{w}) = Z'_P R'_Q(\mathbf{w}) \prod_{i=1}^{n+m} e^{\gamma'_i k(\mathbf{w}|i)}$$

14.4 A Dynamic Language Model: Triggers

We propose to take cognizance of text generation's dynamic character by identifying certain words v of the vocabulary as triggers [3] [4] whose presence in the history $\mathbf{h}$ modifies significantly (i.e., increases or decreases) the distribution of the predicted words w.

Let the trigger words be denoted by $t_1, t_2, \ldots, t_M$, and define indicator functions (bits)

$$b_i(\mathbf{h}) \doteq \begin{cases} 1 & \text{if } t_i \in \mathbf{h} \\ 0 & \text{otherwise} \end{cases} \qquad i = 1, \ldots, M \tag{6}$$

If the fundamental (static) language model is based on trigrams,[13] we can define *equivalent history* by

$$\mathbf{h}^* \doteq w_{-1}, w_{-2}, b_1(\mathbf{h}), \ldots, b_M(\mathbf{h}) \tag{7}$$

and use maximum entropy methods to estimate the probability $P(w_0|\mathbf{h}^*)$.

To do so, we will impose constraints based on the relative frequency of appearance of trigger words in history and *target* words w_0 they trigger in the predicted position. Our task will be first to determine[14] from data the appropriate list of trigger words $t_1, t_2, \ldots, t_M$, and their corresponding target word sets

$$v_1(t_i), v_2(t_i), \ldots, v_{N_i}(t_i) \qquad \text{for } i = 1, 2, \ldots, M$$

and then to compute the distribution $P(w_0|\mathbf{h}^*)$ of the modified language model.

---

13. This method can be made to apply to any other equivalence classification of history.

14. How to make this determination is discussed following equation (9).

Once the trigger words are determined, the desired $P(w_0|\mathbf{h}^*)$ will be that maximal entropy distribution satisfying the following constraints:

$$P(\mathbf{h}^*) = f(\mathbf{h}^*)$$

$$P(w_0, w_{-1}, w_{-2}) = f(w_0, w_{-1}, w_{-2}) \quad \text{if } C(w_0, w_{-1}, w_{-2}) \geq K$$

$$P(w_0, w_{-1}) = f(w_0, w_{-1}) \quad\quad\quad \text{if } C(w_0, w_{-1}) \geq K$$

$$P(w_0) = f(w_0) \quad\quad\quad\quad\quad\quad \text{if } C(w_0) \geq I$$

$$P(w_0 = v_j(t_i), b_i(\mathbf{h}))$$
$$\quad = f(w_0 = v_j(t_i), b_i(\mathbf{h})) \quad\quad \text{for } i = 1, \ldots, M; j = 1, \ldots, N_i;$$
$$b_i(\mathbf{h}) = 0, 1 \text{ provided}$$
$$C(w_0 = v_j(t_i), b_i(\mathbf{h})) \geq L \quad\quad (8)$$

where the relative frequencies f and counts C are obtained from the training data. Note that the last set of constraints involves both 0 and 1 values of $b_i(\mathbf{h})$ because we want to take into account the effect of the presence as well as of the absence of target words.

It may actually be wise to add further constraints involving the frequency with which different trigger words co-occur in the history, since different triggers may predict essentially the same future. In that case we would add to the set (8) constraints of the form

$$P(b_i(\mathbf{h}), b_j(\mathbf{h})) = f(b_i(\mathbf{H}), b_j(\mathbf{h})) \quad\quad\quad\quad\quad\quad\quad\quad (9)$$

We must now show how to determine the trigger words and their targets. Our first inclination would be to pick word pairs with a high mutual information

$$\log \frac{f(w_0 = v, t \in \mathbf{h})}{f(w_0 = v)f(t \in \mathbf{h})} \quad\quad\quad\quad\quad\quad\quad\quad\quad (10)$$

However, (10) would not measure, for instance, how much the absence of the trigger t from the history influences the appearance of v in the predicted position. So instead, we will define a new indicator function

$$b_w(v) \doteq \begin{cases} 1 & \text{if } w = v \\ 0 & \text{otherwise} \end{cases}$$

and another one similar to (6)

$$b_{\mathbf{h}}(t) \doteq \begin{cases} 1 & \text{if } t \in \mathbf{h} \\ 0 & \text{otherwise} \end{cases}$$

and with their help, for each pair t, v, the somewhat modified *average empirical mutual information* function

$I(b_w(v); b_\mathbf{h}(t))$

$$\doteq \begin{cases} \sum_{b_w, b_\mathbf{h}} f(b_w(v), b_\mathbf{h}(t)) \log \dfrac{f(b_w(v), b_\mathbf{h}(t))}{f(b_w(v)) f(b_\mathbf{h}(t))} \\ \qquad\qquad\qquad\qquad \text{if } C(b_w(v) = 1, b_\mathbf{h}(t) = 1) \geq K \\ 0 \qquad\qquad\qquad\quad \text{otherwise} \end{cases}$$

where the sum is over the four possible combinations of values of $b_w(v), b_\mathbf{h}(t)$.

We will then select those trigger-target pairs t, v for which the average empirical mutual information value above exceeds some appropriately chosen threshold.[15]

14.5 The Cache Language Model

As section 14.1 pointed out, the trigger method of making the language model dynamic requires precomputation. In particular, it cannot deal with heretofore unobserved words. If these occur, the language model adjustment must be direct, based on word frequencies observed in the current document so far. This leads to a different type of a dynamic language model, called the *cache language model*, that is particularly suited to text dictation. Its basic idea is simple [5]:

Accumulate word n-grams dictated so far in the current document, and use these to create a local trigram language model $P_c(w_0|w_{-1}, w_{-2})$ which will be interpolated with the precomputed static language model $P_s(w_0|\mathbf{h})$.

Thus we find suitable interpolation weights γ_i and define

$$P_c(w_0|w_{-1}, w_{-2}) = \gamma_3 f_c(w_0|w_{-1}, w_{-2}) + \gamma_2 f_c(w_0|w_{-1}) + \gamma_1 f_c(w_0) \qquad (11)$$

where f_c denotes the relative frequency of word presence in the document dictated so far. The complete language model will then be

$$P(w_0|\mathbf{h}) = (1 - \gamma_c) P_s(w_0|\mathbf{h}) + \lambda_c P_c(w_0|w_{-1}, w_{-2})$$

15. We must restrict the final number M of triggers to limit the amount of computation necessary to determine the parameters defining the desired language model probability $P(w_0|\mathbf{h})$.

where λ_c can be made to vary with the size of the cache (and so can the weights γ_i). Obviously, any words new to the vocabulary will be provided for by the term $P_c(w_0|w_{-1}, w_{-2})$.[16]

Note that while the values of γ_2 and γ_3 would normally be kept quite small, the cache bigram and trigram relative frequencies have the potentially beneficent effect of allowing the language model to predict words belonging to fixed phrases such as *General Electric, Bill Clinton, nuclear magnetic resonance*, etc.

We can further refine formula (11) by taking into account the probability $P(w_0 \in \mathbf{h})$ that the word w_0 will have been observed in the cache at run time.[17] This probability is a function of $\mathbf{h}$ and would be preestimated on training data. The more refined formula would then be[18]

$$P_c(w_0|w_{-1}, w_{-2}) = P(w_0 \notin \mathbf{h})P_s(w_0|\mathbf{h}) + P(w_0 \in \mathbf{h})[\gamma_3 f_c(w_0|w_{-1}, w_{-2})$$
$$+ \gamma_2 f_c(w_0|w_{-1}) + \gamma_1 f_c(w_0)] \tag{12}$$

Additional improvement is possible via the maximum entropy approach [1]. As in section 14.2, we take the original training data as domain #1 and the contents of the cache as domain #2. We estimate a bigram language model $P_c^*(w_0|w_{-1})$ using the following constraints:

$$P_c^*(w_0) = \begin{cases} P(w_0 \in \mathbf{h}) f_c(w_0) & \text{for } w_0 \in \mathbf{h} \\ P(w_0 \notin \mathbf{h}) \dfrac{f_s(w_0)}{\sum_{v \notin \mathbf{h}} f_s(v)} & \text{otherwise} \end{cases}$$

$$P_c^*(w_{-1}) = f_c(w_{-1})$$

The resulting probability will have the form

$$P_c^*(w_0, w_{-1}) = P_s(w_0, w_{-1})g_1(w_0)g_2(w_{-1})$$

16. The idea is that the new words are found through correction of a dictated document. When a word is used for the first time, the recognizer will, of course, misidentify it. The user will correct it, and as a result the recognizer will include it in its (expanded) vocabulary. The relative frequencies f_c are then computed over this dynamically increasing vocabulary and are adjusted continually as recognition and correction proceed.

17. This refinement is desirable because of the empirical observation of users' tendency to repeat words that have particular relevance to their document.

18. Note that if $w_0 \in \mathbf{h}$, then w_0 is in the cache, and if it is not, then the coefficient of $P(w_0 \in \mathbf{h})$ in formula (12) is equal to 0.

The probability $P_c^*(w_0|w_{-1})$ can be used either singly or as a replacement of the unigram term $f_c(w_0)$ in (11). Of course, the factors g_1 and g_2 should be recalculated from time to time on the basis of data currently present in the cache. This might constitute a computational burden considered not worth the potential gain.

14.6   Additional Reading

A very thorough application of the maximum entropy approach can be found in reference [6], which is aimed at language models for statistical machine translation. It can be thought of as a fundamental continuation of the work in [1].

The original results related to cache language modeling discussed in section 14.5 can be found in [7] and [8]. Reference [9] presents another type of language model adaptation at recognition time.

References

[1] S.A. Della Pietra, V.J. Della Pietra, R.L. Mercer, and S. Roukos, "Adaptive language modeling using minimum discriminant information," *Proceedings of the IEEE International Conference on Acoustics, Speech, and Signal Processing*, vol. I, pp. 633–36, San Francisco, 1992.

[2] I.J. Good, "The population frequencies of species and the estimation of population parameters," *Biometrika*, vol. 40, nos. 3 and 4, pp. 237–64, 1953.

[3] R. Lau, R. Rosenfeld, and S. Roukos, "Trigger-based language models: a maximum entropy approach," *Proceedings of the IEEE International Conference on Acoustics, Speech, and Signal Processing*, vol. II, pp. 45–48, Minneapolis, 1993.

[4] R. Rosenfeld, "A maximum entropy approach to adaptive statistical language modeling," *Computer Speech and Language*, vol. 10, no. 3, pp. 187–228, July 1996.

[5] F. Jelinek, B. Merialdo, S. Roukos, and M. Strauss, "A dynamic language model for speech recognition," *Proceedings of Speech and Natural Language DARPA Workshop*, pp. 293–95, Pacific Grove, CA,1991.

[6] A.L. Berger, S.A. Della Pietra, and V.J. Della Pietra,"A maximum entropy approach to natural language processing," *Computational Linguistics*, vol. 22, no. 1, pp. 39–72, March 1996.

[7] R. Kuhn and R. DeMori, "A cache-based natural language model for speech recognition," *IEEE Transactions on Pattern Analysis and Machine Intelligence*, vol. 12, no. 3, pp. 570–83, June 1990.

[8] R. Kuhn and R. DeMori, "Corrections to 'A cache-based natural language model for speech recognition,'" *IEEE Transactions on Pattern Analysis and Machine Intelligence*, vol. 14, no. 3, pp, 691–92, June 1992.

[9] R. Kneser and V. Steinbiss, "On the dynamic adaptation of stochastic language models," *Proceedings of the IEEE International Conference on Acoustics, Speech, and Signal Processing*, vol. II, pp. 586–89, Minneapolis, 1993.

Chapter 15

Estimation of Probabilities from Counts and the Back-Off Method

15.1 Inadequacy of Relative Frequency Estimates

In all previous chapters we have tacitly assumed that relative frequencies of events are a good estimate of their probability. The well-known law of large numbers assures us that this is true to some extent. In particular, if $\mathscr{X}$ is the set of possible events in each of N repeated identically distributed trials, and $C_N(x)$ is the number of times the event $x \in \mathscr{X}$ was observed, then [1]

$$P\left\{ \lim_{N \to \infty} \frac{C_N(x)}{N} = p(x) \right\} = 1 \qquad \text{for all } x \in \mathscr{X} \tag{1}$$

where $p(x)$ denotes the probability that any given trial results in the event x.

Since our definition of relative frequency $f(x)$ is the ratio $C_N(x)/N$, it might seem that our method of estimating $p(x)$ is the best possible. But we have at least three problems:

1. The estimate $C_N(x)/N$ for any particular (and finite) amount N of training data may not be good enough.
2. In many situations of interest to us, the event space $\mathscr{X}$ is not known in its totality.[1]
3. Even when $\mathscr{X}$ is known, it may be so huge[2] that for many $x \in \mathscr{X}$, the true probability will satisfy $0 < p(x) \ll 1/N$, so that the fact that x was not observed in the training data does not indicate at all that its probability should be 0.

---

1. The set of events that took place in the training data may be only a subset of $\mathscr{X}$.
2. For instance, $\mathscr{X}$ can denote the set of trigrams $w_1 w_2 w_3$ taken from a large vocabulary.

Other questions come to mind. Problem 3 above forces us to ask not only whether an unobserved event should have 0 probability, but how much larger a probability should be assigned to an event observed once than to one not observed at all, or, in general, whether the ratio of probabilities of events observed n and m times, respectively, should really be n/m.

In this chapter we will first estimate the desired probabilities by employing our trusted held-out data approach. A variant of that approach will lead to the well-known Good-Turing formulas [2]. We will then show how to enhance these estimation methods by utilizing any side information we might have about the distribution of the process [3]. We will conclude the chapter by deriving the method of back-off estimation [4], which constitutes the basis of most state-of-the-art language models.

15.2 Estimation of Probabilities from Counts Using Held-Out Data

15.2.1 The Basic Idea

Our strategy will be to divide training data into two parts [5]. The first part $\mathscr{D}$, called *development (kept) data*, is going to be used for the collection of counts $C_d(x)$ of events x. The second part $\mathscr{H}$, called *held-out data*, will be used to estimate additional parameters that will determine our estimates $\hat{p}(x)$ of the probabilities $p(x)$ we are after.

We will want $\hat{p}(x)$ to have the following structure:

$$\hat{p}(x) = \begin{cases} q_i & \text{for all } x \text{ for which } C_d(x) = i, \text{ and } i = 0, 1, \ldots, M \\ \alpha f_d(x) & \text{for } C_d(x) > M, \text{ with } f_d(x) \doteq \dfrac{C_d(x)}{N_d} \end{cases} \quad (2)$$

where N_d is the size of the development set $\mathscr{D}$.

We will use the held-out data to find the optimal values of the parameters q_i and α while satisfying the normalization

$$\sum_{x \in \mathscr{X}} \hat{p}(x) = 1 \quad (3)$$

The basic intuition under structure (2) is that all events observed the same number of times $i \in \{0, 1, \ldots, M\}$ should have the same probability. The second case $C_d(x) > M$ represents smoothing[3] and recognizes the following fact:

_____

3. See section 15.3.

Let $R = \max_x C_d(x)$ and assume that M was chosen so that $x \in \mathcal{X}$ exists such that $C_d(x) = M$. Then certain integers j may well exist in the set $\{M + 1, M + 2, \ldots, R - 1\}$ such that no x exists in the development data for which $C_d(x) = j$. Let r be the minimal such integer. Then had we chosen M in (2) to equal R, the parameter estimation procedure we are about to develop would result in $q_r = 0, q_{r-1} > 0$, and $q_R > 0$ and thus would assign a lower probability to an event observed r times than to one observed $r - 1$ times.[4]

Let us denote by n_i the number of different symbols $x \in \mathcal{X}$ for which $C_d(x) = i$. That is,[5]

$$n_i = \sum_x \delta(C_d(x), i)$$

We can give a class-based interpretation to the formula (2). Let x belong to class Φ_i if $C_d(x) = i, i = 0, 1, \ldots, M$, and let x belong to Φ_{M+1} if $C_d(x) > M$. Then

$$\hat{p}(x) = P(\Phi(x))P(x|\Phi(x)) \tag{4}$$

where we have determined beforehand that

$$P(x|\Phi(x)) = \begin{cases} \dfrac{1}{n_i} & \text{if } \Phi(x) = \Phi_i, i = 0, 1, \ldots, M \\[2ex] \dfrac{C_d(x)}{\sum_{j>M} jn_j} & \text{if } \Phi(x) = \Phi_{M+1} \end{cases} \tag{5}$$

and we will use the held-out data to estimate the occupancy probabilities $P(\Phi_i)$. Of course, $q_i = P(\Phi_i)/n_i$ and $\alpha = P(\Phi_{M+1})N_d / \sum_{j>M} jn_j$.

15.2.2 The Estimation
To determine the maximum likelihood values of q_i and α, we need additional notation. We will denote by

---

4. We intend the estimates q_i and α to have a certain universality that would be damaged if $q_r = 0$ and $q_{r+l} > 0$ for some $l > 0$. This will become clear in section 15.3. We will discuss at the end of *this* section how to choose an appropriate value for M.

5. The expression $\delta(a, b)$ denotes the Kronecker delta function

$$\delta(a, b) = \begin{cases} 1 & \text{if } a = b \\ 0 & \text{otherwise} \end{cases}$$

- $C_h(x)$, the number of times x has been seen in $\mathcal{H}$
- r_i, the number of times we have seen in $\mathcal{H}$ symbols x such that $C_d(x) = i$. That is,

$$r_i = \sum_x C_h(x)\,\delta(C_d(x), i)$$

- r^*, the number of times $\mathcal{H}$ contains symbols x such that $C_d(x) > M$. That is,

$$r^* = \sum_{i > M} r_i$$

- $|\mathcal{H}|$, the total size of set $\mathcal{H}$. That is,

$$|\mathcal{H}| = \sum_{i=0}^{M} r_i + r^* \tag{6}$$

The probability of the data in the held-out set $\mathcal{H}$ calculated by (2) is then equal to

$$P(\mathcal{H}) = \prod_{i=0}^{M} (q_i)^{r_i} \times \alpha^{r^*} \times K \tag{7}$$

where

$$K \doteq \prod_{x:C_d(x) > M} f_d(x)^{C_h(x)} \tag{8}$$

We will now choose those values of q_i and α that maximize (7) while satisfying the constraint (3). As before, we will employ the method of undetermined Lagrangian multipliers. That is, we will look for the solution of the following set of equations:

$$\frac{\partial}{\partial q_j}\left[P(\mathcal{H}) - \lambda\left(\sum_{i=1}^{M} n_i q_i + \alpha P_M\right)\right] = 0 \qquad j = 0, 1, \ldots, M$$

$$\frac{\partial}{\partial \alpha}\left[P(\mathcal{H}) - \lambda\left(\sum_{i=1}^{M} n_i q_i + \alpha P_M\right)\right] = 0 \tag{9}$$

$$\sum_{i=0}^{M} n_i q_i + \alpha P_M = 1$$

where

$$P_M \doteq \sum_{x:C_d(x)>M} f_d(x) = \frac{1}{N_d} \sum_{i>M} i n_i$$

Carrying out the derivatives we get

$$\frac{r_j}{q_j} P(\mathcal{H}) = \lambda n_j \qquad j = 0, 1, \ldots, M$$

$$\frac{r^*}{\alpha} P(\mathcal{H}) = \lambda P_M$$

and substituting these relations into (9) we end up with

$$\frac{P(\mathcal{H})}{\lambda} \left(\sum_{i=0}^{M} r_i + r^* \right) = 1$$

Using definition (6), the final solution then is

$$q_j = \frac{1}{n_j} \frac{r_j}{|\mathcal{H}|} \qquad j = 0, 1, \ldots, M \tag{10}$$

$$\alpha = \frac{1}{P_M} \frac{r^*}{|\mathcal{H}|} \tag{11}$$

It is worth noting in particular that

$$q_0 = \frac{1}{n_0} \frac{r_0}{|\mathcal{H}|} > 0$$

and that

$$\frac{q_{j+1}}{q_j} = \frac{r_{j+1}}{r_j} \frac{n_j}{n_{j+1}} \qquad j = 0, 1, \ldots, M - 1 \tag{12}$$

which does not equal $(j+1)/j$, the value one would get if the probabilities were estimated by relative frequency.

15.2.3 Deciding the Value of M

Clearly, we want $\hat{p}(x)$ to be an increasing function of the count $C_d(x)$. So at the very least, we must choose M so that $q_j < q_{j+1}$ for $j = 0, 1, \ldots, M$. This means that M should be chosen so that

$$r_j n_{j+1} < r_{j+1} n_j \qquad \text{for } j = 0, 1, \ldots, M - 1$$

and

$$\frac{r_M}{n_M} < \frac{\sum_{j>M} T_j}{\sum_{j>M} jn_j}(M+1)$$

In fact, if things were to be really smooth, one would like to have[6]

$$\frac{q_{M+1}}{q_M} \cong \frac{(M+1)}{M}$$

Using the values of (10) and (11), this means that we would like to choose M so that

$$\frac{r_M}{Mn_M} \cong \frac{\sum_{j>M} r_j}{\sum_{j>M} jn_j}$$

15.3   Universality of the Held-Out Estimate

We would like the just-obtained estimates of (2) with values (10) and (11) to have universal applicability for the type of data on which the training set is based. To achieve this, we need only to make use of the class interpretation (4) and (5), that is, to recast formula (2) as follows:

$$\hat{p}(x) = \begin{cases} \lambda_i \dfrac{1}{n_i} & \text{for all } x \text{ for which } C(x) = i, \\ & \text{and } i = 0, 1, \ldots, M \\ \lambda_{M+1} \dfrac{C(x)}{\sum_{i>M} in_i} & \text{for } C(x) > M \end{cases} \tag{13}$$

In this way, the λ_i coefficients become *weights* in a deleted interpolation scheme.[7]

Comparing (2) with (13) we get the following relationships

$$\lambda_i = q_i n_i^d = \frac{r_i}{|\mathcal{H}|} \quad i = 0, 1, \ldots, M \tag{14}$$

$$\lambda_{M+1} = \alpha \frac{\sum_{i>M} in_i^d}{N_d} = \frac{r^*}{|\mathcal{H}|} \tag{15}$$

where we use the diacritic d in n_i^d and N_d to indicate that we are dealing with quantities corresponding to a *particular* development set $\mathcal{D}$.

---

6. Since $q_{M+2}/q_{M+1} = (M+2)/(M+1)$.

7. See section 4.3 and 4.4.

If the development plus held-out data were to constitute the total training data available, the final estimates $\hat{p}(x)$ could be improved by calculating them by formula (13), using counts $C(x)$ from the entire corpus, not just from its development part.[8]

We can now return to the topic of section 15.2.3 and of the second paragraph following (3). As we just pointed out, the parameter values λ_i are chosen once and for all on the basis of the development data, but they are to be used permanently for any data of this type. So the fact that for some integer $r < R$ there exists no $x \in \mathcal{D}$ for which $C_d(x) = r$ does not mean that in another training set containing the same type of data we would not find elements x for which $C(x) = r$.

The case just mentioned is, of course, extreme. The reason for assigning $\hat{p}(x) = (\lambda_{M+1})(C(x)/\sum_{i>M} in_i)$ is simply smoothing. It reflects our confidence that for low counts $C(x) \leq M$ (for which the relative sizes of estimated probabilities are crucial) the development set behaved as it should have and leaves the relative distribution of the high counts to the particular data we may encounter in another application.

15.4   The Good-Turing Estimate

One particular formula for estimating probabilities from training counts does not depend on a division into development and held-out sets. It is referred to as the Good-Turing estimate [2] and can be derived by a variety of methods. Nadas showed [6] that the following possible derivation is closely related to the approach of section 15.3.

Let us create a set of pseudo–held-out sets by the "leave-one-out" method. That is, we create N different development sets $\mathcal{D}_j$ and held-out sets $\mathcal{H}_j$ from the training set $\mathcal{T}$ of size N by simply omitting each of the elements x_j of $\mathcal{T} = \{x_1, x_2, \ldots, x_N\}$, $x_j \in \mathcal{X}$, in turn. Thus our development–held-out pairs become

$$\mathcal{D}_j = \mathcal{T} - x_j; \quad \mathcal{H}_j = \{x_j\} \qquad j = 1, 2, \ldots, N$$

We will predict $\mathcal{H}_j$ on the basis of estimates from $\mathcal{D}_j$, and we will choose our parameters so as to maximize the product

---

8. That is, we would first use counts $C_d(x)$ from the development data to calculate the λ_is, and once these were fixed, recalculate $\hat{p}(x)$ according to formula (13).

$$\prod_{j=1}^{N} P_j(\mathcal{H}_j) = \prod_{j=1}^{N} P_j(x_j)$$

where $P_j(x_j)$ denotes the probability of x_j based on the counts obtained from the development set $\mathcal{D}_j$.

Let $C(x)$ and n_i refer to counts gathered from $\mathcal{T}$,[9] let $C_j(x)$ denote the count of x in $\mathcal{D}_j$, and let r_i denote the number of times we have seen, in the totality of held-out sets $\mathcal{H}_j$, $j = 1, 2, \ldots, N$, symbols x such that $C_j(x) = i$. Then

$$r_i = (i+1)n_{i+1} \tag{16}$$

In fact, if $\mathcal{H}_j = x$, then $C_j(x) = C(x) - 1$. So if $C_j(x) = i$ then $C(x) = i + 1$. Furthermore, there are n_{i+1} such elements x in $\mathcal{T}$, and therefore there are $(i+1)n_{i+1}$ different values of $j \in \{1, 2, \ldots, N\}$ for which $\mathcal{H}_j = x$ such that $C_j(x) = i$. This proves (16).

Since in this case $|\mathcal{H}| = N$, we get from (10) that

$$q_j = \frac{n_{j+1}}{n_j} \frac{j+1}{N} \qquad j = 0, 1, \ldots, M \tag{17}$$

We will get the value of α by normalization $\sum \hat{p}(x) = 1$, that is setting,

$$\sum_{j=0}^{M} q_j n_j + \alpha \sum_{j>M} \frac{j}{N} n_j = 1$$

Using (17) we get

$$\alpha = \frac{\sum_{j>M+1} j n_j}{\sum_{j>M} j n_j} \tag{18}$$

We have thus derived in (17) and (18) the well-known Good-Turing formulas.[10]

9. The value of n_i is equal to the number of different elements x for which $C(x) = i$.

10. The natural monotonicity constraint $q_{j-t} < q_j$ now requires the choice of M to satisfy

$$(n_j)^2 < \frac{j+1}{j} n_{j-1} n_{j+1} \quad \text{for } j = 1, 2, \ldots, M$$

and

$$\frac{n_{M+1}}{n_M} N < \frac{\sum_{j>M+1} j n_j}{\sum_{j>M} j n_j}$$

Let us now call attention to the fact that although the occurrence numbers n_j for $j = 1, 2, \ldots, M + 1$ are those actually observed in the data $\mathcal{T}$, n_0 is different. It is an inferred number equal to the size of the total lexicon $\mathcal{X}$ (that we presumably know) minus the size of the sublexicon $\mathcal{X}_T$ actually observed in $\mathcal{T}$.[11] That is,

$$n_0 = |\mathcal{X}| - \sum_{i=1}^{\infty} n_i$$

It is interesting to note from (17) that the total probability $q_0 n_0$ assigned by the formulas to all the unobserved events is equal to n_1/N, that is, the total probability that would have been assigned to singleton events by a relative frequency formula.

We observe from (18) that compared to a frequency allocation formula, the Good-Turing method discounts the high frequency events by the total amount

$$\frac{(M + 1)n_{M+1}}{\sum_{j>M} jn_j}$$

Indeed, the way unobserved events are provided with probability is through a sequence of relative frequency shifts from higher to lower counts.

Finally, since M is usually a small to moderate integer, we call the reader's attention to the formula

$$\sum_{j>M} jn_j = N - \sum_{i=1}^{M} in_i$$

which makes for an easy computation of the left-hand sum.

15.5 Applicability of the Held-Out and Good-Turing Estimates

From more usual proofs (really heuristic arguments) [2] [6] of the Good-Turing estimate, it follows that it is intended for situations in which $n_0 \gg n_1 \gg n_2$, that is, for cases where there are many unobserved events

11. So, for instance, if the lexicon $\mathcal{X}$ is the set of all trigrams $w_1 w_2 w_2$ with words belonging to a known vocabulary $\mathcal{V}$, then $|\mathcal{X}| = |\mathcal{V}|^3$, whereas $\mathcal{X}_T$ simply consists of all the different trigrams actually observed in the training data $\mathcal{T}$.

and the data is really sparse. In fact, the method's original application was estimating the number of unknown biological species. Obviously, formula (17) would not be applicable for $j = 0$ in cases where all elements of $\mathcal{X}$ have been seen in the training data, where $n_0 = 0$.

On the other hand, the held-out estimator is always applicable.[12] Note that if $n_0 = 0$, for instance, then necessarily also $r_i = 0$ so that (14) remains valid. This estimate's obvious disadvantage is that it is not a ready-made formula, that the procedure must be carried out separately for every different domain. This would present particular difficulty for estimating distributions at the leaves of decision trees as these trees are being developed. In fact, leaf splitting depends on the distribution at the leaf. So we would need two held-out sets: one for estimating those distributions and another for imposing the stopping rule.

If the held-out estimator is always valid, why isn't Good-Turing, which we just derived from the held-out? The reason is that our argument tacitly assumed a certain regularity in our observation of events. Events observed once had to be very rare and could have just as well not have been observed at all. Events observed $j + 1$ times could have just as well been observed only j times, and this situation was well represented by the leave-one-out computation. That can be the case only for huge event spaces where the events are, so to speak, anonymous. In general, the frequency of occurrence of one event has nothing to say about the frequency of another.

In trigram language models we are interested in probabilities $P(w_3|w_1, w_2)$. There is an obvious temptation to use the Good-Turing estimate directly, that is, to use counts $n_i(w_1, w_2)$ of the number of different words w_3 that have followed the bigram $w_1 w_2$ exactly i times in the training data.[13] This temptation must be resisted because in most cases the counts would be based on far too little data.

In trigram language models, what is needed is the computation

$$P(w_3|w_1, w_2) = \frac{P(w_1, w_2, w_3)}{\sum_{w_3'} P(w_1, w_2, w_3')} \tag{19}$$

12. That does not mean that it cannot be misused, for instance by setting aside an excessively small held-out set or having the wrong balance between the development and held-out sets.

13. In this case, $n_i(w_1, w_2)$ and $n_j(w_1', w_2')$ would be determined completely independently of each other.

Table 15.1

Multiplicity n_j	Rel. freq. $C(x) = j$	Held-out $\frac{1}{n_j} \frac{r_j}{|\mathcal{H}|} N_d$	Good-Turing $\frac{n_{j+1}}{n_j}(j+1)$
74,671,100,000	0	0.000027	0.000027
2,018,046	1	0.448	0.446
449,721	2	1.25	1.26
188,933	3	2.24	2.24
105,668	4	3.23	3.24
68,379	5	4.21	4.22
48,190	6	5.23	5.19

in which the probabilities $P(w_1, w_2, w_3)$ are the result of Good-Turing estimation. Formula (19) should, of course, also be used when the estimate is based on the held-out method of section 15.2. Formula (19) is quite feasible computationally, since the denominator can be precomputed.[14]

Table 15.1 compares bigram probabilities estimated by the relative frequency, held-out, and Good-Turing methods as computed by Church and Gale [3] from a training sample of size 22,000,000. The comparison is made in terms of the quantity $N\hat{p}(w_1, w_2)$, which enables us to see what a single count is worth. It is interesting that for this example (only), after count 2, each additional count is worth a contribution of approximately a single count to relative frequency (see the table's last two columns). At the same time, the first count is worth the contribution of only about half of a count, and the second count only about 0.8 counts. This would indicate that instead of formula (2), we should, at least in this case, use a formula like

14. The calculation of $\sum_{w_3'} P(w_1, w_2, w_3')$ is in any case easy. Let $m_i(w_1, w_2)$ denote the number of trigrams w_1, w_2, w_3' that have been observed i times in the data. Then

$$\sum_{w_3'} P(w_1, w_2, w_3') = \sum_{i=0}^{M} q_i\, m_i(w_1, w_2) + \frac{\alpha}{N} \sum_{i=M+1}^{\infty} i\, m_i(w_1, w_2)$$

where q_i and α are given by (10) and (11), respectively, and in the Good-Turing case by (17) and (18).

$$\hat{p}(x) = \begin{cases} q_i & \text{for all } x \text{ for which } C_d(x) = i, \text{ and } i = 0, 1, \ldots, M \\ \dfrac{\beta + C_d(x)}{N_d} & \text{for } C_d(x) > M, \text{ with } f_d(x) \doteq \dfrac{C_d(x)}{N_d} \end{cases}$$

and estimate parameters q_i and β for it.

15.6 Enhancing Estimation Methods

Both our estimation methods have as their basis the structure (2) which says that the probability of each element of $\mathscr{X}$ should be determined by its frequency in the development set. But suppose we have some additional side information which could allow us to distinguish between elements having the same frequency? Church and Gale [3] pointed out how such information could be used.

15.6.1 Frequency Enhancement of Held-Out Estimation of Bigrams

Suppose the lexicon $\mathscr{X}$ consists of bigrams $w_1 w_2$. Is there any reason to believe that two different bigrams, neither of which is observed, should be assigned a different probability? Here is a possible argument: If both w_1 and w_2 are frequent, then it is most probably not just bad luck that the bigram $w_1 w_2$ has not been observed. On the other hand, if they are infrequent, then their nonoccurrence may not signify any allergy between them.

So one way to get a better estimate is to categorize each bigram $w_1 w_2$ not just by its frequency of occurrence, but also by its product probability $\hat{p}(w_1)\hat{p}(w_2)$.[15] Obviously, it will be impossible to base categories on the exact value of $\hat{p}(w_1)\hat{p}(w_2)$—we will have to partition the interval $[0, 1]$ into subintervals and classify $w_1 w_2$ by the interval into which $\hat{p}(w_1)\hat{p}(w_2)$ falls. Church and Gale recommend that these subintervals be logarithmic and that their exact size be determined so that the corresponding class (subinterval) contains a sufficient number of different bigrams. It is intuitively obvious that low counts $C(w_1, w_2)$ should be divided into more $\hat{p}(w_1)\hat{p}(w_2)$–determined subclasses that high counts.

To find the estimation formula, we must examine the argument of section 15.2 that resulted in the estimates (10) and (11). We will find that these remain unchanged if we reinterpret our notation as follows:

15. Normally we would use the estimate $\hat{p}(w) = f(w)$, taking the view that $f(w)$ is based on enough data to make it an accurate estimate of probability.

$$\hat{p}(x) = \begin{cases} q_i & \text{for all } x \text{ belonging to class } i \in \{1, 2, \ldots, L\} \\ \alpha f_d(x) & \text{for } x \text{ belonging to class } L+1, \text{ where } f_d(x) \doteq \dfrac{C_d(x)}{N_d} \end{cases}$$

(20)

With the following definitions,

• $\Phi(x)$, the function mapping x into one of the categories $\{1, 2, \ldots, L+1\}$.

• n_i, the number of different symbols $x \in \mathcal{X}$ that belong to category i. That is,

$$n_i = \sum_x \delta(\Phi(x), i)$$

• $C_h(x)$, the number of times x has been observed in $\mathcal{H}$

• r_i, the number of times we have seen in $\mathcal{H}$ symbols x such that $\Phi(x) = i$. That is,

$$r_i = \sum_x C_h(x) \delta(\Phi(x), i) \quad i = 1, 2, \ldots, L+1$$

• $|\mathcal{H}|$, the total size of set $\mathcal{H}$. That is,

$$|\mathcal{H}| = \sum_{i=1}^{L+1} r_i$$

• P_L, the development set probability of the $L+1^{th}$ category. That is,

$$P_L = \sum_{x:\Phi(x)=L+1} f_d(x)$$

the optimal held-out estimate remains

$$q_j = \frac{1}{n_j} \frac{r_j}{|\mathcal{H}|} \quad j = 1, 2, \ldots, L$$

$$\alpha = \frac{1}{P_L} \frac{r_{L+1}}{|\mathcal{H}|}$$

and therefore its universal form (14) and (15) stays valid as well.

15.6.2   Frequency Enhancement of Good-Turing Estimation of Bigrams

The enhancement idea applies also to the Good-Turing estimates because they were derived from (10) and (11). However, we will have to modify our notation somewhat.

- Let $\varphi_{i,j}, j = 1, 2, \ldots, J_i$ be a partition of $\mathscr{X}$ intended to serve for the classification of events x seen i times in training. That is,

$$\mathscr{X} = \bigcup_j \varphi_{i,j} \quad \text{and} \quad \varphi_{i,j} \bigcap \varphi_{i,l} = \phi \text{ for } j \neq L$$

- Let Ψ_i be the set of events $x \in \mathscr{X}$ such that $C(x) = i$.
- Let $n(\Psi_i, \varphi_{l,j})$ be the number of distinct events $x \in \Psi_i \bigcap \varphi_{l,j}$.
- Let $n_i = \sum_{j=1}^{J_i} n(\Psi_i, \varphi_{i,j})$.

Then, assuming that counts $M + 1$ or larger have only one class (i.e., $J_{M+1} = 1$ and $\varphi_{M+1,1} = \mathscr{X}$), we get the enhanced Good-Turing estimate

$$\hat{p}(x) = \begin{cases} \dfrac{n(\Psi_{i+1}, \varphi_{i,j})}{n(\Psi_i, \varphi_{i,j})} \dfrac{i+1}{N} & \text{for all } x \in \Psi_i \bigcap \varphi_{i,j}, i \in \{0, 1, \ldots, M\}, \\ & j \in \{1, 2, \ldots, J_i\} \\ \dfrac{\sum_{j>M+1} jn_j}{\sum_{j>M} jn_j} f(x) & \text{for } x \text{ such that } C(x) > M, \text{ where } f(x) = \dfrac{C(x)}{N} \end{cases}$$

15.6.3 Other Enhancements

Obviously, many enhancements are possible, as seen from the formulation at the end of section 15.6.1. The basic idea simply is to subcategorize the data on the basis of observations of the development set. For instance, when estimating trigrams, one can subcategorize by the value of $\hat{p}(w_1)\hat{p}(w_2)\hat{p}(w_3)$ or even by $\hat{p}(w_1, w_2)\hat{p}(w_3|w_2)$, where the probabilities used for subcategorization are themselves enhanced estimates.

Another obvious possibility is to subcategorize on part-of-speech sequences as follows. Map every word $v \in \mathscr{V}$ into its most likely part of speech $g(v)$. Then subcategorize low-count bigrams $w_1 w_2$ by their part-of-speech membership $g(w_1)g(w_2)$. In this way unseen bigrams DET DET will have a different probability from bigrams NOUN NOUN, and both will differ from the probability of bigrams DET NOUN.

Finally, assuming that the desired language model prediction is of the form $\hat{p}(w_3|w_1, w_2)$, one could subcategorize on the basis of $f(w_2, w_3)$ provided it is not equal to 0, and if it is, then on the basis of $f(w_3)$. The result would be something akin to the usual back-off estimate [4]

$$\hat{p}(w_3|w_1, w_2) = \begin{cases} f(w_3|w_1, w_2) & \text{if } C(w_1, w_2, w_3) \geq K \\ \alpha f(w_3|w_2) & \text{if } C(w_2, w_3) \geq L \\ \beta f(w_3) & \text{otherwise} \end{cases}$$

15.7 The Back-Off Language Model

As we first pointed out in chapter 4 (and many times since), any estimation of probabilities for a language model that depends on history[16] necessarily suffers from the sparse data problem. Until now we have attempted to alleviate that problem by linear interpolation. So for trigrams, we defined the language model probabilities by

$$P(w_3|w_1, w_2) = \lambda_3 f(w_3|w_1, w_2) + \lambda_2 f(w_3|w_2) + \lambda_1 f(w_3) \tag{21}$$

where we estimated the weights λ_i from held-out data. However, Katz argued [4] that if the count $C(w_1, w_2)$ is sufficiently large, then $f(w_3|w_1, w_2)$ by itself is a much better estimate of $P(w_3|w_1, w_2)$ than the one given above, and that a form such as

$$\hat{p}(w_3|w_1, w_2) = \begin{cases} f(w_3|w_1, w_2) & \text{if } C(w_1, w_2, w_3) \geq K \\ \alpha Q_T(w_3|w_1, w_2) & \text{if } 1 \leq C(w_1, w_2, w_3) < K \\ \beta \hat{P}(w_3|w_2) & \text{otherwise} \end{cases} \tag{22}$$

would constitute a superior approximation, provided the factors (not necessarily constant) α and β were appropriately chosen.[17] In (22), $Q_T(w_3|w_1, w_2)$ is a Good-Turing–type function (see below) and $\hat{P}(w_3|w_2)$ is a bigram probability estimate having the same form as $\hat{P}(w_3|w_1, w_2)$:

$$\hat{P}(w_3|w_2) = \begin{cases} f(w_3|w_2) & \text{if } C(w_2, w_3) \geq K \\ \alpha Q_T(w_3|w_2) & \text{if } 1 \leq C(w_2, w_3) < K \\ \beta f(w_3) & \text{otherwise} \end{cases} \tag{23}$$

Form (22) then constitutes a recursion.[18]

How do we determine the functions Q_T? If the threshold K were infinite, then Katz would suggest $\alpha = 1$ and

$$Q_T(w_3|w_1, w_2) \doteq \frac{P_T(w_1, w_2, w_3)}{f(w_1, w_2)} \tag{24}$$

---

16. That is, it is not based just on unigrams.

17. Among other requirements, they must assure proper normalization of the probability $\hat{P}(w_3|w_1, w_2)$.

18. In both (21) and (22) it is tacitly assumed that $f(w_1, w_2) > 0$. If this is not so, then in (21) $\lambda_3 = 0$, whereas (22) is simply not applicable and $\hat{P}(w_3|w_1, w_2) \doteq \hat{P}(w_3|w_2)$, the latter being defined by (23). Note that we always take it for granted that each vocabulary word has been seen in the training data, so $f(w_3) > 0$ for all $w_3 \in \mathscr{V}$.

where the numerator is the Good-Turing probability obtained from the collection of counts $C(w_1, w_2, w_3)$ by the formulas of section 15.4. The definition (24) of Q_T does not constitute a probability because this function is not properly normalized, which it would be only if $f(w_1, w_2)$ were replaced by $\sum_{w_3} P_T(w_1, w_2, w_3)$. The formula (24) is simple and allows for an easy determination of the factors α and β. We will deal with the normalization of the resulting $\hat{P}(w_3 | w_1, w_2)$ at the end of our development.

Since the threshold K is finite (Katz advocates $K \cong 6$), things are a little more complicated. We want to determine α so as to satisfy the following basic principle:

The total probability allocated to all unseen trigrams[19] should be n_1/N (as prescribed by Good-Turing for $K = \infty$), which will determine the value of α. We have

$$n_1 = \sum_{w_1, w_2, w_3} \delta(C(w_1, w_2, w_3), 1)$$

and

$$N = \sum_{w_1, w_2, w_3} C(w_1, w_2, w_3)$$

Now by formula (17),

$$P_T(w_1, w_2, w_3) = \frac{n_{r+1}}{n_r} \frac{r+1}{N} \qquad \text{where } r = C(w_1, w_2, w_3)$$

and

$$f(w_1, w_2, w_3) = \frac{r}{N} \qquad \text{where } r = C(w_1, w_2, w_3)$$

so the above principle will be satisfied if

$$\sum_{r=K}^{\infty} r n_r + \alpha \sum_{r=1}^{K-1} (r+1) n_{r+1} = N - n_1 = \sum_{r=2}^{\infty} r n_r$$

Therefore,

$$\alpha = \frac{\sum_{r=2}^{K-1} r n_r}{\sum_{r=2}^{K} r n_r} \tag{25}$$

19. We *do* mean the probability of trigrams, *not* the conditional probability of the third word given the first two.

Finally, for $\hat{P}(w_3|w_1, w_2)$ to be normalized, $\beta = \beta(w_1, w_2)$ must satisfy

$$\beta \sum_{w_3 \in \mathscr{S}_0} \hat{P}(w_3|w_2) = 1 - \sum_{w_3 \in \mathscr{S}_K} f(w_3|w_1, w_2) - \alpha \sum_{w_3 \in \mathscr{S}^*} Q_T(w_3|w_1, w_2)$$

(26)

where $\mathscr{S}_0 = \{w_3 : C(w_1, w_2, w_3) = 0\}$, $\mathscr{S}_K = \{w_3 : C(w_1, w_2, w_3) \geq K\}$, and $\mathscr{S}^* = \{w_3 : 1 \leq C(w_1, w_2, w_3) < K\}$. We will not derive the formulas for α, β, and $Q_T(w_3|w_2)$ appropriate for (24), as the required development is identical to the one we just discussed.

15.8 Additional Reading

Three articles, all coauthored by Hermann Ney, study additional aspects of backing-off. The first [7] establishes a relationship between linear interpolation (e.g., formula (21)) and backing-off, and then considers various methods of *discounting*[20] for backing-off. In particular, the article treats *absolute discounting* in which a fixed amount D is subtracted from the count $C(x)$ of each observed event x. Many other problems of language modeling are also considered from a general point of view, such as generalized cache language modeling, smoothing with grammatical and other equivalence classes, and the like.

The second article [8] is similar in content to the first, but it also contains other results and concentrates on the leave-one-out method that we used in section 15.4 to derive the Good-Turing estimate. Finally, the third article [9] improves on the backing-off formula (22) by replacing the distribution $\hat{P}(w_3|w_2)$ with a new distribution $Q(w_3|w_2)$ such that the resulting $\hat{P}(w_3|w_1, w_2)$ distribution becomes consistent with $\hat{P}(w_3|w_2)$, that is that

$$\hat{P}(w_3|w_2) = \sum_{w_1} \hat{P}(w_3|w_1, w_2)\hat{P}(w_1|w_2)$$

is satisfied. In this case, of course, we must estimate α, β, and $Q(w_3|w_2)$, and the authors show us how.[21]

20. This standard term refers to the fact that in the Good-Turing formula, counts of observed events are discounted to provide probability mass for unobserved events. So, an event observed r times is assigned the count $((r+1)n_{r+1})/n_r$ instead, and the difference between these two counts, $(rn_r - (r+1)n_{r+1})/n_r$ is called the discount.

21. See the notation of section 15.7.

There are other methods of assigning probabilities from counts that in one way or another account for unseen events. Most famous among them is Laplace's law of succession [10] [11], which adds count 1 to each observation,

$$P(v) = \frac{C(v) + 1}{\sum_{v'} C(v') + |\mathscr{V}|}$$

Other laws can be found in the book by Lehmann [11] and in the literature of data compression. A very interesting formula has recently been derived by Ristad [12].

References

[1] A.F. Karr, *Probability*, pp. 187–90, Springer Verlag, New York, 1993.

[2] I.J. Good, "The population frequencies of species and the estimation of population parameters," *Biometrika*, vol. 40, parts 3 and 4, pp. 237–64, December 1953.

[3] K.W. Church and W.A. Gale, "A comparison of the enhanced Good-Turing and deleted estimation methods for estimating probabilities of English bigrams," *Computer Speech and Language*, vol. 5, pp. 19–54, 1991.

[4] S. Katz, "Estimation of probabilities from sparse data for the language model component of a speech recognizer," *IEEE Transactions on Acoustics, Speech and Signal Processing*, vol. 35, no. 3, pp. 400–01, March 1987.

[5] F. Jelinek and R.L. Mercer, "Probability distribution from sparse data," *IBM Technical Disclosure Bulletin*, vol. 28, pp. 2591–94, 1985.

[6] A. Nadas, "Good, Jelinek, Mercer, and Robbins on Turing's estimate of probabilities," *American Journal of Mathematical and Management Sciences*, vol. 11, nos. 3/4, pp. 229–308, 1991.

[7] H. Ney, U. Essen, and R. Kneser, "On structuring probabilistic dependencies in stochastic language modeling," *Computer Speech and Language*, vol. 8, no. 1, pp. 1–38, January 1994.

[8] H. Ney, U. Essen, and R. Kneser, "On the estimation of 'small' probabilities by leaving one out," *IEEE Transactions on Pattern Analysis and Machine Intelligence*, vol. PAMI-17, no. 12, pp. 1202–12, December 1995.

[9] K. Kneser and H. Ney, "Improved backing-off for *m*-gram language modeling," *Proceedings of the IEEE International Conference on Acoustics, Speech, and Signal Processing*, pp. 181–84, Detroit, MI, May 1995.

[10] P.S. Laplace, *Philosophical Essay on Probabilities*, Springer Verlag, New York, 1995.

[11] E.L. Lehmann, *Theory of Point Estimation*, John Wiley, New York, 1983.

[12] E.S. Ristad, *A Natural Law of Succession*, Research Report CS-TR-495-95, Dept. of Computer Science, Princeton University, Princeton, NJ, 1995.

Name Index

Subject Index

Bold numbers indicate pages on which term is defined or discussed in depth.